Log Evaluation of Shaly Sandstone Reservoirs: A Practical Guide

GEORGE B. ASQUITH

A NOTE TO THE READER

The intent of this text is to provide you with a simplified review of shaly sand well log analysis. Voluminous literature exists on the subject and the intent of this work is not to replicate it. Rather, I hope you will find this work useful as a quick look, ready reference guide. Because of the important effects that shale or clay have on both reservoir quality and well log response, a working knowledge of the subject provides a necessary basis for accurate log analysis. Techniques illustrated in the text help you make better decisions about the productive potential of shaly reservoirs and also help you more effectively map shaly sandstones. Shaly sand techniques are presented for old (pre-porosity) logs and for modern logging suites. Both "old" and "new" techniques are needed in order to maximize the use of existing log data. Many of the logs we work with are older logs and often even newer wells are logged with an incomplete log suite. This text demonstrates that although many older methods have limitations, they can still be surprisingly effective in an evaluation of shaly reservoirs.

Two invaluable publications on shaly sand well log analysis for those who want to pursue a more in-depth study are: The Shaly Sand Reprint Volume, published by the Society of Professional Well Log Analysts, edited by W. H. Lang; and Log Analysis of Sand-Shale Sequences—A Systematic Approach, by Poupon, et al.

George B. Asquith
Adobe Chair of Petroleum Geology
Texas Tech University
Lubbock, Texas

TABLE OF CONTENTS

ABOUT THE AUTHOR

George B. Asquith—Dr. Asquith is presently the Adobe Chair of Petroleum Geology, Department of Geosciences at Texas Tech University, Lubbock, Texas. Dr. Asquith received his Ph.D. from the University of Wisconsin in 1966, and since that time has worked for ARCO Research, Search Drilling Company, Pioneer Production Company, and Mesa Limited Partnership. He also taught previously at West Texas State University. Besides being a member of the AAPG, Dr. Asquith is a member of the SPWLA, SEPM, AIPG, and has served as Vice President and President of the PGS. He has written more than 30 articles and abstracts on carbonate and sandstone petrology, sedimentation, and computer geology. He has been involved in several AAPG training programs and publications, including the *Handbook of Log Evaluation Techniques For Carbonate Reservoirs* (AAPG Methods in Exploration Series, #5), *Basic Well Log Analysis For Geologists* (AAPG Methods in Exploration Series, #3), and the present new course note: *Log Evaluation of Shaly Sandstones: A Practical Guide*. He is involved in two short courses: *How To Evaluate Carbonate Reservoirs From Well Logs*, and *Formation Evaluation of Shaly Sand Reservoirs* (from which his new course note has evolved). He is also involved in a school, *Well Log Analysis*, as the Science Advisor. The AAPG is appreciative of Dr. Asquith's contributions over the years to the dissemination of information to geologists.

CHAPTER 1

INTRODUCTION

In any petroleum province in the world where there are sand/shale sequences, the geologist, engineer or log analyst confronts the "shaly sand problem." The shaly sand problem is really a question of identifying, from logs, the degree to which the presence of clay degrades reservoir quality, because clay affects both porosity and permeability. If too much shale or clay is in a reservoir, production of hydrocarbons is suppressed by excessive reduction in permeability. However, a small amount of clay disseminated in the pores can be helpful for trapping interstitial water, permitting hydrocarbon production from reservoirs with high water saturation.

Most hydrocarbon-bearing sandstones contain laminated or dispersed clay minerals (Fertl, 1982). As a consequence, any assessment of subsurface strata or reservoir (unless it has been conclusively established otherwise) should begin with the assumption that clay may be affecting log-derived measurements of porosity and water saturation.

An understanding of the nature of clay and of the log data corrections needed if significant amounts of clay are present in a reservoir helps the dedicated learner avoid serious misjudgments in both hydrocarbon development and exploration work. While it is true that the mathematical formulas and adjustments necessary to obtain the correct information from logs in shaly sand reservoirs are laborious and time-consuming, the end result of better formation evaluation is more than worth the effort.

Computers are an invaluable tool for performing the complicated formulas of shaly sand analysis, but their use does require a thorough knowledge of shaly sand models. Otherwise, producibility versus nonproducibility will be confused, volumetric calculations may be incorrect, and serious completion problems will be unanticipated.

In part, the ubiquitousness of clay in sandstones is strongly related to the number of depositional environments with which clay is associated. For example, shaly (clayey) sands are common in deep-water turbidite and mass-flow deposits where clay particles aid in the transport of the coarser fraction. In shallow-water deltaic environments delta front and prodelta sands are deposited interlaminated with shales and clays. Along wave-dominated coasts, bioturbated shoreface sands have shales and clays organically reworked and mixed with sand. Offshore or marine bar-sands often have a lower energy shaly sand facies associated with winnowed higher-energy clean sands.

Another, and, in terms of log effect, even more important source of clay minerals found in sandstones is the dispersed, authigenic clays formed either as a direct precipitate from formation waters or through the diagenetic alteration of feldspars and volcanic rock fragments. These authigenic clays may be very common. For example, Wilson and Tillman (1974) in an analysis of 785 sandstone samples from the Tertiary to Paleozoic, found authigenic clay in 90 percent of the samples they studied.

As the major component of shale, clay minerals consist of extremely fine-grained hydrous alumino-silicates. They have several characteristics which can directly affect logs. Clay minerals exhibit varying degrees of radioactivity because of unstable isotopes in their lattice. Density may also vary depending on the type of clay mineral. However, it is the characteristics of clay affecting electrical behavior that are the most significant to log interpretation. One such characteristic is the charge imbalance occurring along clay surfaces, allowing exchange of cations (cation exchange capacity or CEC) between the clay's adsorbed water and formation water.

An additional clay characteristic is the large surface area of the clay minerals (Berry, 1982). As a result of having a large surface area (Table 1), clay is able to adsorb large quantities of a reservoir's pore water to its surface. This bound water contributes to electrical conductivity (i.e. lowers resistivity on resistivity logs)*, but does not contribute to hydraulic conductivity because the bound water is immovable. A reservoir with dispersed clay will have both effective porosity (non-clay bound fluid) and total porosity (clay bound and non-clay bound fluid).

The presence of shale or clay in a sandstone has two effects on the reservoir: 1) shale or clay reduces a reservoir's storage capacity by reducing effective porosity, and 2) shale or clay reduces a reservoir's ability to transmit fluids by reducing permeability. In addition to the loss of effective porosity and permeability, clay can also cause problems when a well is completed. These problems include: 1) migration of fines—where clay minerals are torn loose from the sand grains and migrate with the reservoir's fluids to the well bore and plug pores, causing a loss in permeability; 2) water sensitivity—where the clays swell in the reservoir when frac water is introduced, causing both a loss of effective porosity and permeability; and 3) acid sensitivity—where hydrochloric acid (HCl) reacts with iron-bearing clays to form ferric hydroxide, a gelatinous precipitate that clogs pores and reduces permeability.

In Table 2, Ali (1981) summarizes some of the reservoir problems associated with each group of clay minerals and identifies some of the reservoirs that contain

*Remember: lower resistivity means higher calculated Archie water saturation.

Table 1. Specific Surface Areas of Various Minerals (Measured by Nitrogen Adsorption) (Modified after Almon, 1979).

Quartz	0.15 cm^2/gm*
Kaolinite	23 m^2/gm
Smectite	752 m^2/gm**
Illite	113 m^2/gm
Chlorite	42 m^2/gm

Note: The surface area of clays is orders of magnitude more than the surface area of quartz. This large surface area means that clay is able to bind large quantities of immovable (bound) water to the surface of the clay.
* Depends on grain size and distribution.
** Calculated by N. Guven (1988).

these clays. It becomes obvious after studying Table 2 that knowledge of a reservoir's clay mineralogy must be carefully considered when designing reservoir treatments.

The presence of clay in a reservoir has two effects on petrophysical logs: l) it lowers resistivity (Fig. 1), and 2) it causes the porosity logs (sonic, neutron, and density) to generally record too high a porosity.

Figure 2 is a plot of formation water resistivity (R_w) versus the resistivity of a water saturated sandstone (R_o). A clean (shale-free) sand will plot as a straight line passing through the origin with a slope equal to one over formation factor (F*) times the resistivity of water (R_w). The equation can be stated $1/F \times R_w$. If the clay content of sand increases by replacement of some of the rock matrix, and if the effective porosity remains unchanged, the line in Figure 2 is displaced downward, indicating less resistivity but greater conductivity. This loss of resistivity (Fig. 2) is due to high bound water content and greater conductance associated with the clay in the sandstone. Note in Figure 2 that the clay effect is not uniform and lowers resistivity more at higher formation water resistivities (R_w).

In order to determine true water saturation in a shaly sand, the Archie water saturation equation must be modified by a number of different shaly sand equations which are discussed later in the text. The presence of clay causes the three porosity logs—sonic, density, and neutron—to generally record too high a porosity. The first porosity log listed the sonic log, records too high a porosity because the interval transit time of clay or shale is high. When the sonic porosity is calculated using the reservoir's matrix travel time, the calculated porosity is higher than it should be. The second porosity log listed the neutron log, measures hydrogen ion concentration and records clay as porosity because of its high hydrogen content. As a result, calculated neutron porosity errs by being too high. Finally, the last porosity log in our list, the density log, records too high a porosity whenever the matrix density of the clay is less then the reservoir's matrix density. However, in the case where clay's density and a reservoir's matrix density are the same, the density log will measure true effective porosity. If clay density is greater or less than the matrix density (2.65 g/cc), density log porosities can be less than or greater than the porosity calculated in a clay-free sandstone.

Table 2. Principal Clay Mineral Groups and the Specific Reservoir Problems Caused by Each Group (After Ali, 1981).

GROUP	COMPOSITION	SPECIES	MAJOR RESERVOIR PROBLEMS	EXAMPLES
Kaolinite	$Al_4[Si_4O_{10}]\,(OH)_8$	Kaolinite Dickite Nacrite Halloysite	Migration of fines	Brent Ss Ft .Union Fm Woodbine Fm Morrow Ss
Smectite	$(1/2\ Ca, Na)_{0.7}(Al, Mg, Fe)_4$ $[(Si,Al)_8O_{20}](OH)_4 \bullet nH_2O$	Montmorillonite Beidellite Nontronite Hectorite Saponite	Water sensitivity Microporosity	Second Wall Creek Ss Norphlet Fm Frontier Fm Mesaverde Fm
Illite	$K_{1-1.5}Al_4[Si_{7-6.5}Al_{1-1.5}O_{20}]\,(OH)_4$	Illite	Microporosity Migration of fines	Rotliegendes Ss Wilcox Fm Cotton Valley Ss Brent Ss Muddy Fm
Chlorite	$(Mg,Fe,Al)_{12}[(Si,Al)_8O_{20}](OH)_{16}$	Amesite Greenalite Chamosite Clinochlore Penninite	Acid sensitivity	Frio Fm Shannon Ss Woodbine Fm Hosston Ss Morrow Ss
Mixed-Layer Clay Minerals		Illite/ Smectite Chlorite/ Smectite Chlorite/ Vermiculite	Water sensitivity Microporosity Acid sensitivity	Norphlet Fm Mesaverde Fm Wilcox Fm Guadalupe Fm

*Formation factor is derived from the equation F = a (tortuosity factor)/xm (cementation exponent).

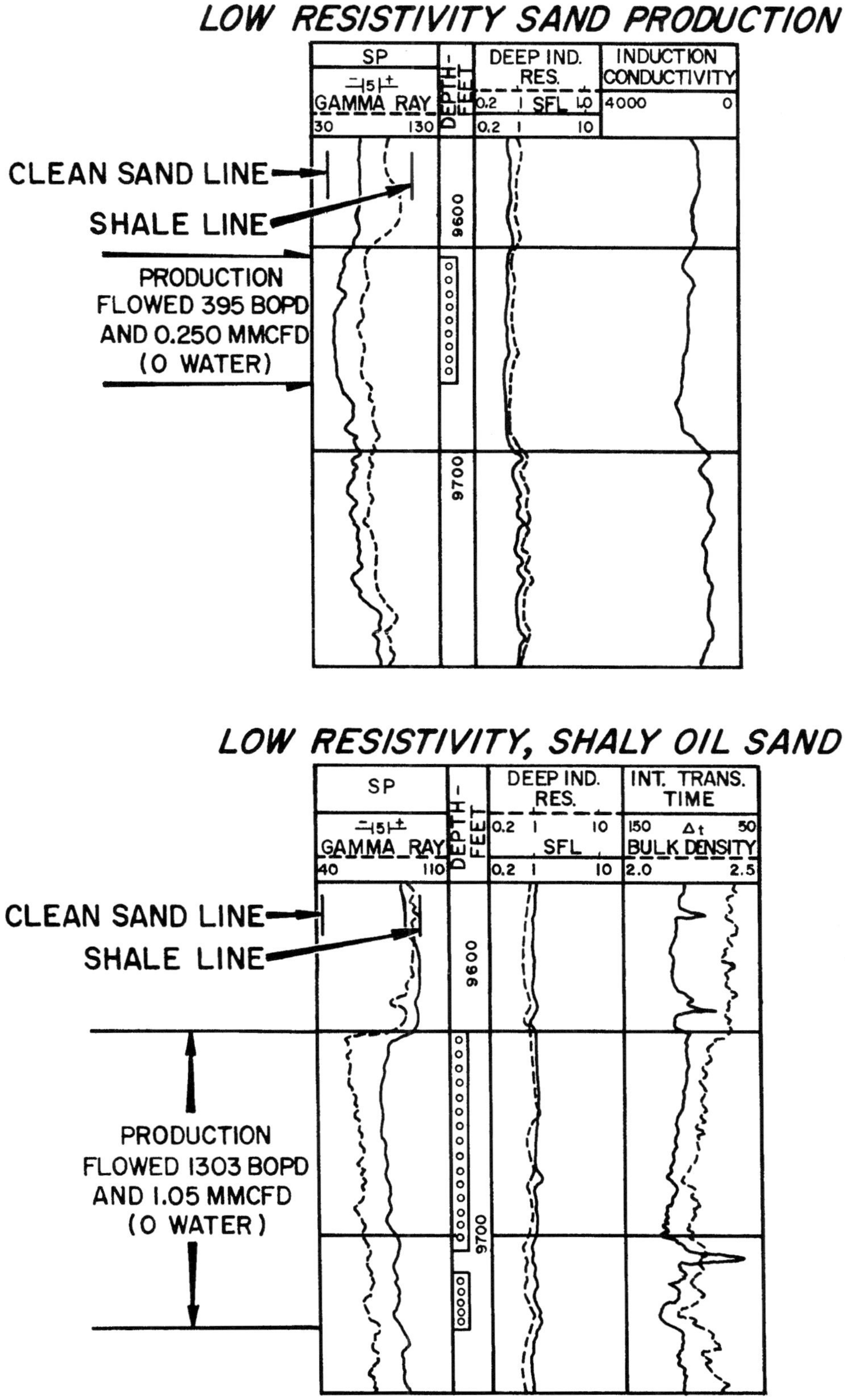

Figure 1—Low resistivity productive Tertiary sands from offshore Gulf of Mexico: A. Sand is clean (low shale content) and consists of very fine-grained sand to coarse silt with minor clay coatings. Note that the gamma ray reading of the sand is close to the clean sand line. B. Example of low resistivity shaly sand. Note that the shaly sand has a much higher gamma ray reading than the clean sand line. Modified after Vajnar, et al., 1977.

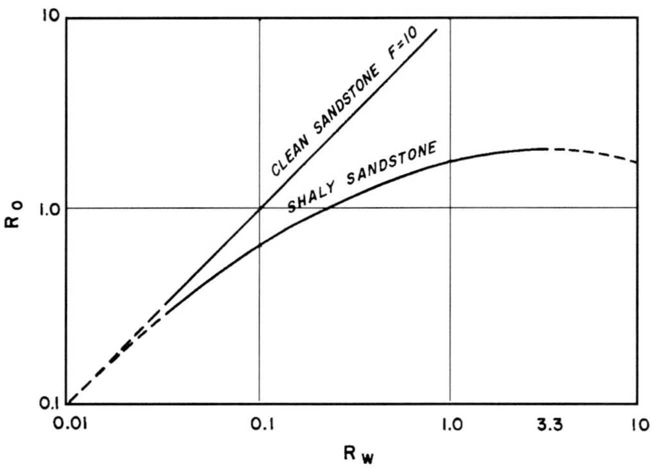

Figure 2—Relationship between resistivity of a porous water-bearing formation and formation water resistivity for clean and shaly sandstones of the same porosity (ϕ = 31.6). The lower curved line (lower resistivities) is the result of the addition of clay to the sandstone. This loss of resistivity, due to the presence of shale or clay, will result in an increase in calculated water saturations (S_w). Note that the response to shale or clay is not uniform; the loss of resistivity is higher at greater R_w values. Modified after Sneider, 1987.

Generally, we can say that in shaly sand analysis, the porosities measured by all three porosity logs must be corrected for the presence of clay to find the actual fluid transmittable or effective (ϕ_e) porosity, for it is only the effective porosity that is capable of allowing movement of reservoir fluids.

Shaly sand analysis requires several steps, including correcting total porosity (ϕ_t; all porosity including both clay-free and clay-bearing) to effective porosity (ϕ_e; only clay-free porosity). In addition, total water saturation (S_{wt}; before shale or clay adjustment) must be corrected to reveal effective water saturation (S_{we}; after shale or clay adjustment). Shale or clay may cause the uncorrected total water saturation (S_{wt}) to be very high (S_{wt} > 70%), indicating that a zone is water saturated (wet), when, in fact, it may be productive. Consequently, shaly sand corrections are designed to discriminate between total water saturation and effective water saturation (i.e. $S_{we} < S_{wt}$).

The importance of shaly sand analysis cannot be overstated. By using the corrections that constitute what is called "shaly sand analysis," we can determine the percentage of effective porosity (ϕ_e) that controls the flow of fluids—oil, gas or water; we can determine effective water saturation (S_{we}) so that potential hydrocarbon producing zones are not overlooked; and we can use shaly sand analysis to help us evaluate potential enhanced recovery (EOR) shaly reservoirs.

There are vast amounts of hydrocarbons tied up in enhanced recovery (EOR) shaly reservoirs (Tyler, 1988) that count as undeveloped or unproduced reserves. These represent an important future oil and gas supply once the hydrocarbons are freed by enhanced oil recovery. A vital first step in the recovery of these reserves is shaly sand analysis, because effective porosity (ϕ_e) and effective water saturation (S_{we}) must be known in order to both map and volumetrically determine the amount of possible recoverable hydrocarbon.

This text is devoted to discussing techniques which help answer the questions: Is shale or clay present in a reservoir? And, if shale or clay is present, how can log readings be adjusted to yield correct information about fluid transmitting porosity and moveable water? Both effective porosity and effective water saturation yield information about reservoir producibility and volumetrics.

Before these questions can be addressed and the actual log analysis of shaly reservoirs presented, it will be helpful if the reader acquires an understanding of the nature of shale, shale and clay distributions, and how shale and clay affect resistivity and porosity logs. With this in mind, we now turn our attention to these subjects.

CHAPTER 2

NATURE OF CLAY MINERALS AND SHALE

General

Shale is a mixture of clay minerals and silt laid down in very low energy environments. The silt fraction* of shale consists of fine particles (<0.0625 mm), mostly quartz. On the other hand, the clay fraction of shale is made up of minerals which are hydrous aluminum silicates with small amounts of magnesium, iron, potassium, and other elements. Clay particles have a layered platelet structure 5–10 Å (Å = 10^{-8} cm) thick and to 10,000 Å in length and width. Clay platelets are stacked one above the other with spacings of 20–100 Å. Clay minerals are very small particles (maximum size 0.0002 mm), about 10–100 times smaller than the average sand grain (Dewan, 1983).

Clay minerals can be classified into specific groups according to their crystal structure. Table 3 is a list of significant clay minerals and illustrates clay properties that affect log readings. The first column in Table 3 gives the cation exchange capacity (CEC) of each clay mineral.

The cation exchange capacity (CEC) is the capacity of the clay to exchange cations from exchangeable positions in the adsorbed water around the clay with cations in the free water in the pores, and with cations in the interlayer of the clay minerals. Because of the higher cation exchange capacity (CEC) of montmorillonite and illite compared with chlorite and kaolinite (Table 3), shaly sands with large amounts of montmorillonite and illite have higher conductivity (i.e. lower resistivity) than shaly sands with kaolinite and chlorite.

In Table 3, chlorite and kaolinite have much greater neutron porosities than montmorillonite and illite. Consequently, shaly sands with large amounts of chlorite and kaolinite have higher neutron porosities when compared with shaly sands composed predominantly of montmorillonite and illite. Calculated density porosity is affected by the presence of montmorillonite and chlorite. The higher density of chlorite (2.8 gm/cc) versus quartz (2.65 gm/cc) results in a lower calculated density porosity. But, the lower density of montmorillonite (2.45 gm/cc) versus quartz (2.65 gm/cc) results in a higher calculated density porosity.

The most important minor component in clays is iron (Fe; Table 3). The presence of iron affects acid sensitivity to hydrochloric acid (HCl). The spectral gamma ray components listed in columns 6-8 (Table 3) cause the high gamma ray readings in shales.

The clay minerals listed in Table 3 can occur in the reservoir as mineralogical mixtures, and when they do, they are called mixed-layered clays.

*Silt fractions with large amounts of potassium feldspar have a high gamma ray response, leading to overestimation of clay content.

Shale or Clay Distribution in Shaly Sands

Laminated Shale

For the purpose of log analysis, the term "laminated shale" refers to strata of shale a fraction of an inch to several inches thick interbedded with sand or sandstone (Fig. 3B). Laminar shales may be so thin that the resistivities of the clean, hydrocarbon-bearing sands are averaged by the resistivity log with the clay-rich siltstones or shales, and produce a shale-like, low resistivity log reading. Because shale has essentially zero permeability, the average permeability and porosity of interlaminated sand/shale sequences will theoretically be reduced in proportion to the volume of shale (V_{sh}). For example, 40 percent laminar shale will theoretically reduce effective permeability and porosity by 40 percent to only 60 percent of the clean sand values. By using a rule of thumb that a 50 percent reduction in permeability is the cut-off for production, we might assume that a reservoir could produce hydrocarbons with as much as 50 percent laminar shale. However, in nature, the laminated sand/shale sequences tend to be lenticular, and consequently, 30 to 40 percent laminated shale is the normal tolerated maximum for commercial production in this type of shale distribution (Dewan, 1983).

The detrital origin of clay in laminated shale means that shale laminations between sands normally have the same clays and water content as the adjacent thick shale beds. This similarity in laminated shales leads to a basic assumption that the resistivities of laminated shale and thick adjacent shales are the same. However, it is necessary to first establish that a reservoir approximates a shaly sand laminar distribution model before we can use the resistivities of adjacent shale beds. Other clay distributions can preclude the use of adjacent bed resistivities.

Structural Shale

When occurring in the form of structural shale, clay grains, shale clasts or particles are present along with sand grains (Fig. 3C). Usually, structural clays occur as grains or clasts and, as such, have little effect on porosity and permeability. This type of clay is least objectionable to reservoir quality, and commonly does not occur in effective quantities. If it is conclusively established that structural shale is present, then, because of its detrital ori-

Table 3. Clay Properties of Concern in Logging. (Modified after Dewan, 1983).

Clay Type	CEC meq/g	dCNL	ρ(av), g/cc	Minor Constituents	Spectral GR Components (av) K, %	U, ppm	Th, ppm
Montmorillonite	0.8-1.5	0.24	2.45	Ca,Mg,Fe	0.16	2-5	14-24
Illite	0.1-0.4	0.24	2.65	K,Mg,Fe,Ti	4.5	1.5	< 2
Chlorite	0-0.1	0.51	2.8	Mg,Fe	—	—	—
Kaolinite	0.03-0.06	0.36	2.65	—	0.42	1.5-3	6-19

CEC - Cation Exchange Capacity
ρ(av) - Average Density
GR - Gamma Ray
Th - Thorium
CNL - Neutron porosity
K - Potassium
U - Uranium

NOTE:

1. Clays with higher CEC result in higher conductivity (i.e. lower resistivity).
2. The presence of chlorite and kaolinite in a shaly sand results in higher total neutron porosity, compared with shaly sands with montmorillonite and illite.
3. Only with montmorillonite and chlorite does the presence of clay greatly affect the density log, because montmorillonite (2.45 g/cc) and chlorite (2.8 g/cc) have different densities than quartz (2.65 g/cc).
4. The presence of iron (Fe) in clays causes acid sensitivity to HCl.
5. The spectral gamma ray components cause high gamma ray readings in shales only if the clay in the shale is montmorillonite, illite, or kaolinite.

Note: Clays with high CEC values (Montmorillonite and Illite) are called effective clays; clays with low CEC values (Chlorite and Kaolinite) are called noneffective clays.

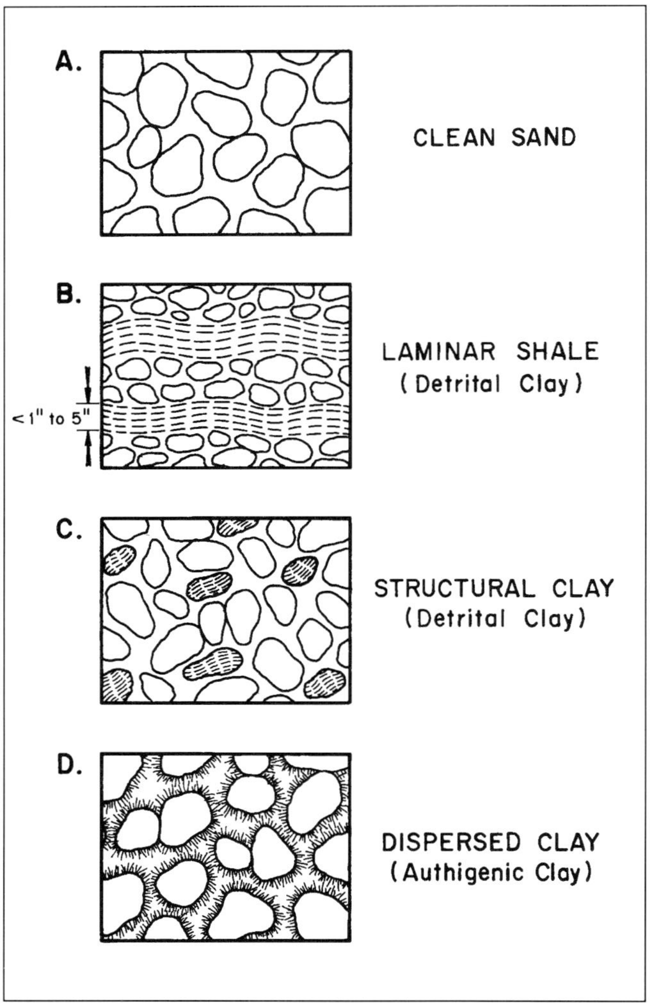

Figure 3—Mode of occurrences of shale and clay in sandstones. A. Clean sand. B. Laminar shale (detrital clay) <l in. to 5 in. lamination thickness. C. Structural (framework) clay (detrital clay) D. Dispersed (pore) clay (authigenic clay).

gin, adjacent shale beds can be used to establish shale resistivity. Structural and laminated shales have similar log responses (Visser, et. al., 1988).

Dispersed Clay

Dispersed clay is disseminated in the pore space of sand and replaces fluid volume (Fig. 3D). This type of clay distribution is very damaging because a small amount of dispersed clay chokes pores and reduces effective porosity and permeability to nonproductive values. Maximum tolerable clay by volume is 15–20 percent. Most of the dispersed clay is authigenic; in other words it diagenetically formed in place after deposition of the sand. The authigenic origin of most dispersed clay means that the composition of the clay (including water content) in the shaly sand and the clay in the adjacent shale may be radically different, and may have significantly different resistivities. Thus, in shaly sand analysis, it is risky to use adjacent shale resistivities, unless it has already been established that the shale distribution is laminated (Fig. 3B), structural (Fig. 3C), or allogenic bioturbated clays. In all these cases the origin is detrital; adjacent shale resistivities are therefore compatible with the shale in the shaly sand, so that shaly sand equations requiring a shale resistivity value can be used. When authigenic dispersed clay is present or suspected, or when clay distribution is unknown, shaly sand equations which do not require a value for shale resistivity (R_{sh}) are needed.

Figure 4 illustrates three different ways dispersed, authigenic clay can be present in a shaly sand. The discrete-particle type (Fig. 4) consists mainly of kaolinite, which builds up isolated platelets that reduce porosity and permeability (Fig. 5) very little. The pore-lining type (Fig. 4) coats the grains with "whiskers" forming micropores that trap pore water and significantly lower permeability (Fig. 5). The pore-bridging type (Fig. 4) chokes

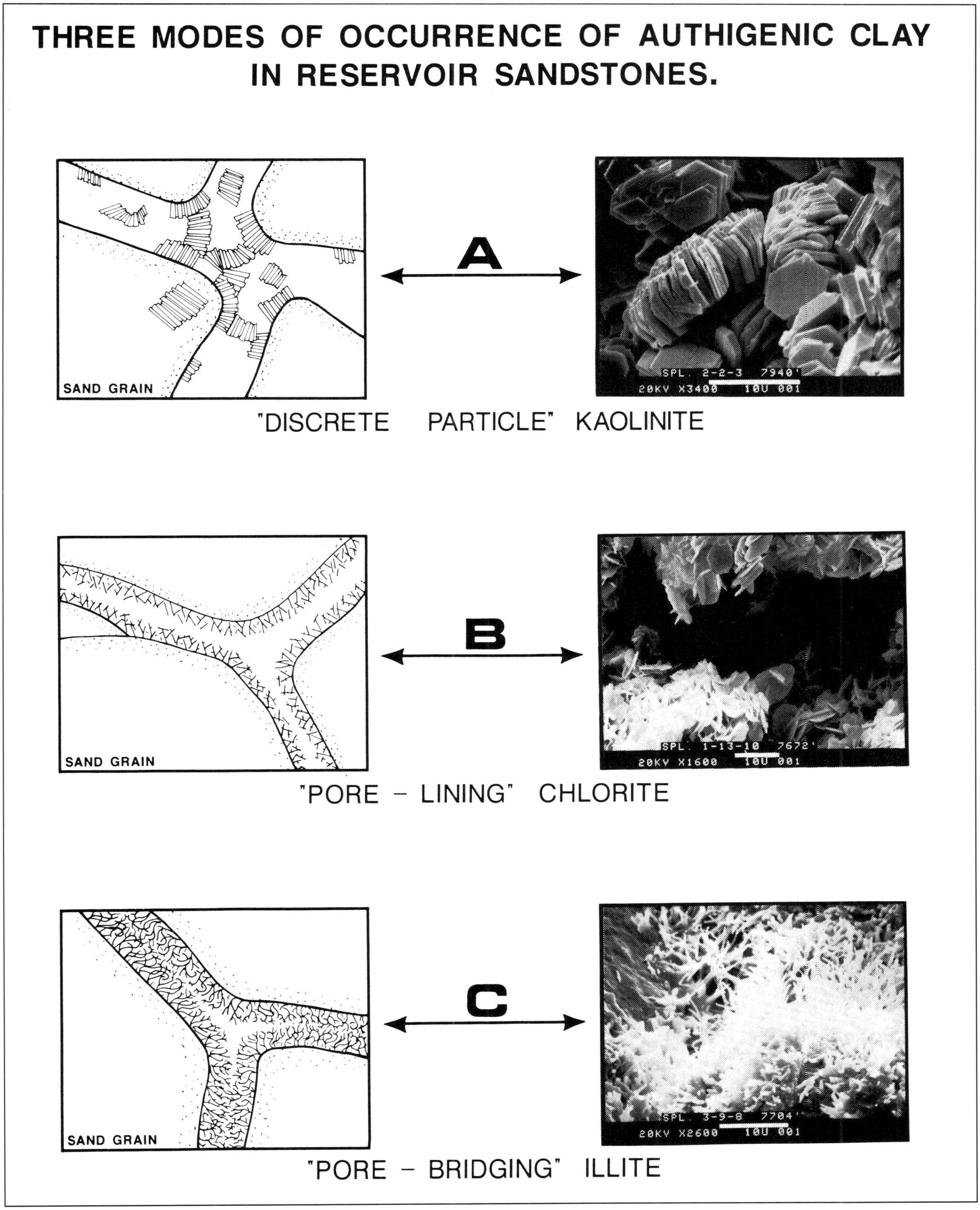

Figure 4—Three modes of occurrence of authigenic clay in reservoir sandstones (after: Munson, 1988). In this Pennsylvanian upper Morrow Sandstone example, the discrete particle clays are kaolinite, the pore-lining clays are chlorite, and the pore-bridging clays are illite.

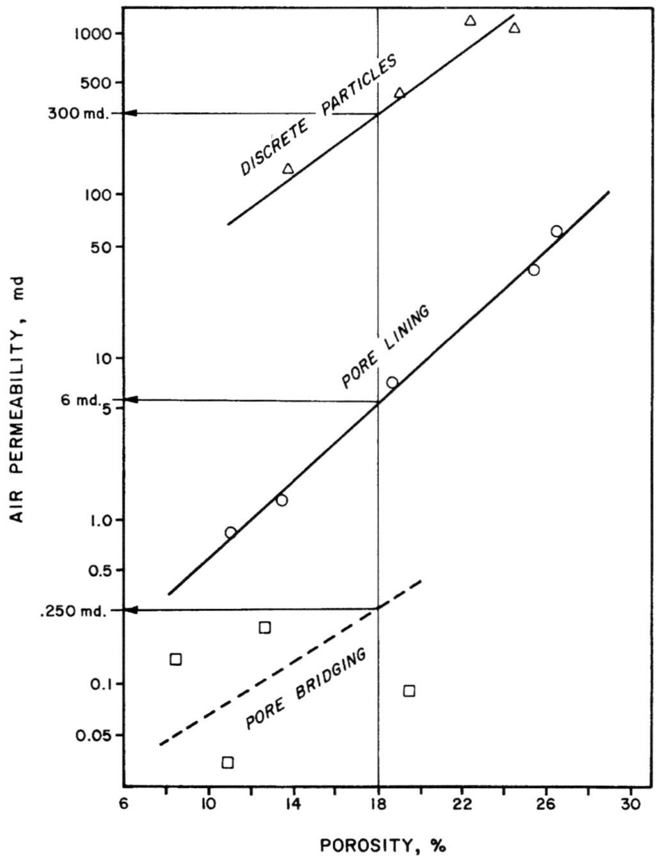

Figure 5—Effect of authigenic pore clays on permeability. Notice that for a constant porosity, the presence of the different types of authigenic clays have dramatically different effects on permeability.

Example: at 18% porosity, the permeabilities for the different types of pore clay are: Discrete particle—300 md. Pore-lining—6 md. Pore-bridging—0.25 md. (Modified after: Neashan, 1977, ©SPE-AIME).

the pores with fibrous clay that significantly lowers effective porosity and drastically reduces permeability (Fig. 5).

Clay Minerals and Resistivity Logs

The presence of clay in a sandstone reservoir lowers the resistivity measured by the induction electric logs (Fig. 1). This occurs because of the large amount of bound water held by very fine-grained clay minerals and also because of the high cation exchange capacity (CEC) of some clays (Table 1). As mentioned earlier, the maximum size of clay minerals is approximately 0.0002 mm, or about 10 to 100 times smaller than the average sand grain. This very fine grain size, along with the often fibrous shape of clays (Fig. 4), results in a shaly sand reservoir which has a large amount of bound water (i.e., high S_{wirr}). The high bound water content of shaly sands results in a lower resistivity but since the bound water is immobile, the lower resistivity may not necessarily mean a water-productive reservoir.

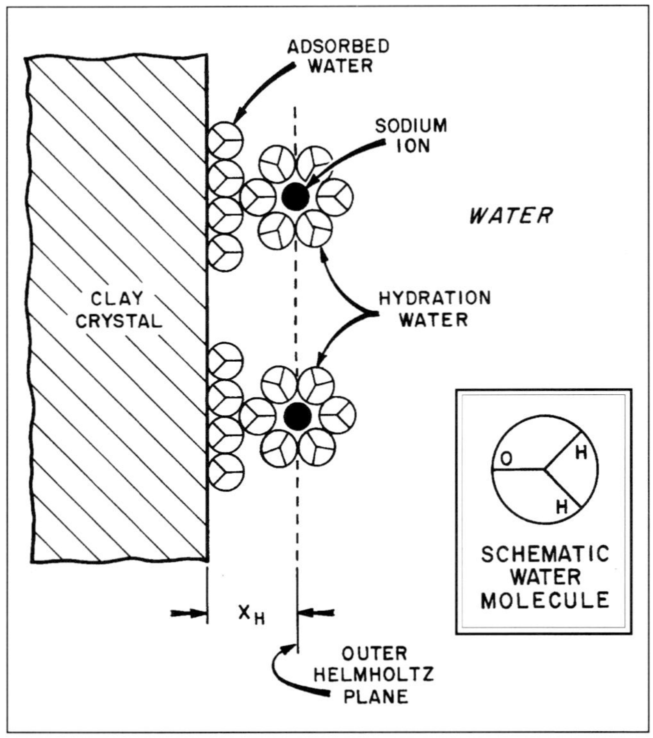

Figure 6—Model of water and distribution of exchangeable cations on the clay surface. The cations in the exchangeable sites in the adsorbed water are called exchangeable cations. The greater the amount of these cations the higher the cation exchange capacity (CEC) and the greater the surface conductance of the clay. (After: Clavier, et. al., 1977. © SPE—AIME.)

Clay platelets are negatively charged due mainly to ionic substitutions in the crystal structure and from broken bonds along their edges. When a clay platelet is in a saline solution, charge-balancing Na^+ ions are held near the clay surface, and Cl^- anions are repelled from the clay surface.

Figure 6 illustrates the distribution of Na^+ ions and H_2O molecules near a clay platelet. Directly on the clay surfaces is a monolayer of adsorbed water. Next is a layer of hydrated Na^+ ions, sufficient to balance the negative clay platelet charge. The Na^+ ions in Figure 6 are called exchangeable cations, and their presence on the surface of the clay platelet causes an increase in surface electrical conductivity (i.e., lower resistivity). Therefore, the greater the CEC* of the clays (Table 3) in the shaly sand, the more the resistivity will be lowered. Figure 7 shows that CEC of the various clay minerals is a function of the clay mineral's specific surface area, and specific surface area is a function of grain size.

We conclude, then, that the presence of clay in a sandstone can lower resistivity, sometimes dramatically, and can result in a very high calculated Archie water saturation if corrections are not made for the effects of clay.

* CEC - cation exchange capacity expressed in milliequivalents per gram of dry clay (1 meq - 6 x 10^{20} atoms). CEC may also be expressed as milliequivalents/unit volume of pore fluid (Q) where: Q = CEC x ρ (1 x ϕ)/ϕ meq/cc. (Dewan, 1983).

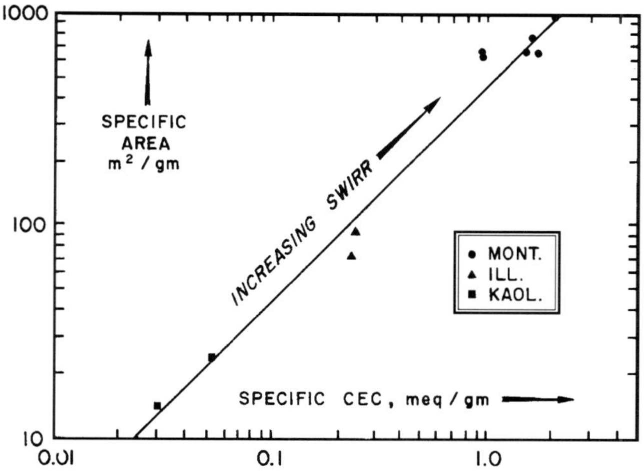

Figure 7—Relationship of specific surface area to cation exchange capacity (CEC). (After: Patchett, 1975.) Along with the increase in specific surface area and CEC, there is a concomitant increase in the reservoir's irreducible water saturation (S_{wirr}), because the greater the specific surface area, the greater the amount of clay-bound water.

Clay Minerals and Porosity Logs

As stated earlier, the presence of clay causes all the porosity logs (neutron, density, and sonic) to record too high a porosity. The only exception to this is the density log, which will not record too high a porosity if the clay density (Table 3) is equal to or greater than the reservoir's matrix density (Quartz = 2.65 g/cc).

The low velocity of clay is recorded on the sonic log by a high interval transit time reading. Then, when sonic porosity of a shaly sand is calculated, the high interval transit time results in a calculated high sonic porosity. The neutron log, which records hydrogen ion concentration, detects the presence of clay in a sandstone as an increase in porosity, because of the large amount of hydrogen in the clay's crystal structure and in the clay's hydrated water. The density log records too high a porosity in a shaly sand when the density of the clay is less than the density of the sand matrix (2.65 gm/cc).

It is vital that the porosity of a shaly sand be corrected for the presence of shale or clay to effective porosity. By definition, only effective porosity transmits the reservoir fluids of oil, gas and water, and is an essential parameter in our pursuit of commercial versus noncommercial hydrocarbon production. Just how we adjust standard log equations to find effective porosity and water saturation is the next focus of our discussion.

CHAPTER 3

SHALY SAND ANALYSIS

General

The basic formulas which make up shaly sand analysis are presented and discussed in this chapter. We begin by introducing methods for determining the volume of clay (V_{cl}) in a reservoir. Once the amount of clay has been estimated, we use the volume of clay (V_{cl}) value to correct data from porosity logs and to calculate effective porosity (ϕ_e).

After identifying the methods used to find volume of clay (V_{cl}) and effective porosity (ϕ_e), this chapter continues with log analysis procedures used to correct data from logs to yield the effective water saturation (S_{we}) of shaly sands. Water saturation formulas are organized around log suites, since the available log package largely dictates which formulas must be used to find the effective (as opposed to total) water saturation. Our reader may need to choose a technique based on either the date of a log or by log suite depending on available information. Time divisions are presented in the text not as an exercise in the history of logging, but rather as a practical guide for rapid identification of the strategies needed to evaluate a particular log.

Time is subdivided by the early 1950s, the late '50s, and the 1960s, 1970s, and 1980s. The reader need only find his or her log dates, check them against the historical breakdown presented here, and against log type, and then use the formulas suggested by the text to find effective water saturation.

We conclude our discussion in this chapter with some helpful guidelines that are applied when doing log analysis of shaly sand reservoirs. Most particularly, we discuss the criteria that are used for selecting the right shaly sand methods.

In all shaly sand analysis, formulas are used to reduce the water saturation value from the value that is calculated if the effects of shale or clay are ignored. Lowering the water saturation numbers can be a problem in log evaluation, because if the volume of clay (V_{cl}) is over-estimated, a water-bearing zone may calculate as if it were a hydrocarbon zone (Hilchie, 1978).

A major problem that still confronts us in shaly sand analysis is determining a shale or clay resistivity value for the shale or clay in shaly sand zones. The percentage of clay in a reservoir is less important than clay type and its cation exchange capacity (Table 3). As stated earlier, in shaly sand analysis, we often make an assumption that the resistivity of the adjacent shale beds is the same as the shale or clay in the reservoir rock. There are exceptions to this rule, and the reader is reminded of the caveats that pertain to use of resistivities of the adjacent shale. Although several cautions have already been touched upon (see discussion of laminated, structural and dispersed shales and clays, Chapter 2), these will be expanded upon as we progress through the text.

Determination of Volume of Clay

The first step in shaly sand analysis is to determine volume of clay (V_{cl}). This is a volumetric number and does not take into account such variables as type of clay mineral and distribution. It is, however, a clay indicator, and so unless the geologist, engineer or log analyst knows that clay or shale is not a reservoir problem in their area, volume of clay (V_{cl}) should always begin any log analysis. Three different logs are used in volume of clay calculations: gamma ray, spontaneous potential and neutron-density. The gamma ray log detects clay by an increase in radiation associated with the presence of clay; the spontaneous (SP) log detects clay by the decrease in SP deflection caused by the loss of ionic permeability in the shaly sand; and the neutron-density log detects the presence of clay by the increase in neutron porosity over density porosity. This large amount of neutron porosity is the result of the high hydrogen-ion concentration in clays.

The gamma ray log equations necessitate knowing whether or not sand in a given area is consolidated or unconsolidated. Commonly, Tertiary sands are unconsolidated.

1. GAMMA RAY LOG

$V_{cl} = 0.33[2^{(2 \times I_{GR})} - 1.0]$ (consolidated sands)

$V_{cl} = 0.083[2^{(3.7 \times I_{GR})} - 1.0]$ (unconsolidated sands)

where:

V_{cl} = volume of clay

$$I_{GR} = \frac{GR_{log} - GR_{min}}{GR_{max} - GR_{min}}$$

I_{GR} = Gamma Ray Index

GR_{log} = gamma ray* from shaly sand

GR_{min} = gamma ray minimum (clean sand)

GR_{max} = gamma ray maximum (shale)

*Whenever possible, use the gamma ray log recorded with the neutron-density porosity tool, because the logging speed of the neutron-density tool is slower than the induction log. Also, if a spectral gamma ray log is available, use its gamma ray minus uranium curve to calculate volume of clay.

2. SP LOG

$$V_{cl} = 1.0 - \frac{PSP}{SSP}$$

where:
V_{cl} = volume of clay
PSP = pseudostatic spontaneous potential (SP in shaly sand)
SSP = static spontaneous potential (SP in thick clean sand)

3. NEUTRON-DENSITY* LOG
$V_{cl} = (\phi_n - \phi_d)/(\phi_{nsh} - \phi_{dsh})$
where:
V_{cl} = volume of clay
ϕ_n = neutron porosity in shaly sand
ϕ_d = density porosity in shaly sand
ϕ_{nsh} = neutron porosity of adjacent shale
ϕ_{dsh} = density porosity of adjacent shale

Volume of clay can be determined by applying all three methods, and then by using the lowest obtained value in the porosity and shaly sand equations, presented later in this chapter. However, if the shaly sand is a gas sand, volume of clay (V_{cl}) should not be determined from the neutron-density log. Gas may not evenly affect the neutron curve or density curve, and so, in this situation, the gamma ray log and SP methods are compared to find the lowest value for volume of clay (V_{cl}).

Even though volume of clay (V_{cl}) can be arrived at by the suggested comparison techniques, the best volume of clay (V_{cl}) value may be the one based on the gamma ray log. If a gamma ray log is available, research by Johnson and Linke (1977) indicates that using the formula provided in the text for finding volume of clay (V_{cl}) from gamma ray logs yields a volume of clay (V_{cl}) value that works as the most effective number. For example, Johnson and Linke (1977) report excellent correlation between volume of clay (V_{cl}) determined using the gamma ray log and cation exchange capacity (CEC) values measured on cores. They find, however, poor correlation between volume of clay (V_{cl}) and cation exchange capacity (CEC) determined from neutron-density logs, and attribute this to the fact that the gamma ray log responds primarily to the higher radioactive content in illite and montmorillonite, which have high cation exchange capacities (CEC). Neutron-density logs, when used as the basis for finding volume of clay (V_{cl}), tend to exaggerate the influence of kaolinite and chlorite, because these clay minerals have higher neutron porosities, compared with those of illite and montmorillonite (Table 3). But, because both kaolinite and chlorite have low cation exchange capacity (CEC) values compared with illite and montmorillonite (Table 3), there is a poor correlation between neutron-density derived volume of clay (V_{cl}) and cation exchange capacity (CEC). This suggests that the gamma ray log is a better indicator of cation exchange capacity (CEC) than are the neutron-density logs. As mentioned before in the text, clays with high cation exchange capacity (CEC) also tend to have the highest surface area (Fig.7); therefore, they also have the highest bound water content, and have a more pronounced effect on log measurements. We conclude, then, that a value for volume of clay (V_{cl}) that favors clays with high cation exchange capacity (CEC) leads to better corrections in the porosity and shaly sand equations.

The gamma ray log should not be used to determine volume of clay when feldspars are present in the sand. The presence of potassium in the feldspars will cause an overestimation of volume of clay.

*Neutron and density porosities must be corrected to the proper matrix.

Correction of Porosity Logs for Volume of Clay

After a volume of clay (V_{cl}) value is determined, this value can be used to correct the porosity log for the effect of clay. The formulas for correcting the sonic, density, and combination neutron-density logs for volume of clay (V_{cl}) are:

$$\phi_e = \frac{\Delta t_{log} - \Delta t_{ma}}{\Delta t_f - \Delta t_{ma}} \times \frac{100}{\Delta t_{sh}} - V_{cl}\left(\frac{\Delta t_{sh} - \Delta t_{ma}}{\Delta t_f - \Delta t_{ma}}\right)$$

Sonic log (Dresser Atlas, 1979)
where:
ϕ_e = effective porosity (sonic log corrected for clay)
Δt_{log} = interval transit time of shaly formation
Δt_{ma} = interval transit time of the formation's matrix
Δt_f = interval transit time of fluid (189 fresh mud and 185 salt mud)
Δt_{sh} = interval transit time of adjacent shale
V_{cl} = volume of clay

$$\phi_e = \frac{\rho_{ma} - \rho_b}{\rho_{ma} - \rho_f} - V_{cl}\left(\frac{\rho_{ma} - \rho_{sh}}{\rho_{ma} - \rho_f}\right)$$

Density log (Dresser Atlas, 1979)
where:
V_{cl} = volume of clay
ϕ_e = effective porosity (density log corrected for clay)
ρ_{ma} = density of the formation's clay-free matrix
ρ_b = density of shaly formation
ρ_f = density of fluid (1.0 fresh mud and 1.1 salt mud)
ρ_{sh} = density of adjacent shale

Combination Neutron-Density log (Dewan, 1983)
$\phi_{nc} = \phi_n - (V_{cl} \times \phi_{nsh})$
$\phi_{dc} = \phi_d - (V_{cl} \times \phi_{dsh})$
$\phi_e = (\phi_{nc} + \phi_{dc})/2$ for oil
$\phi_e = [(\phi_{nc}^2 \times \phi_{dc}^2)/2]^{1/2}$ for gas
where:
ϕ_e = effective porosity (neutron and density logs corrected for clay)
ϕ_n = neutron porosity of shaly formation
ϕ_d = density porosity of shaly formation
V_{cl} = volume of clay
ϕ_{nc} = neutron porosity corrected for clay
ϕ_{dc} = density porosity corrected for clay
ϕ_{nsh} = neutron porosity of shale
ϕ_{dsh} = density porosity of shale

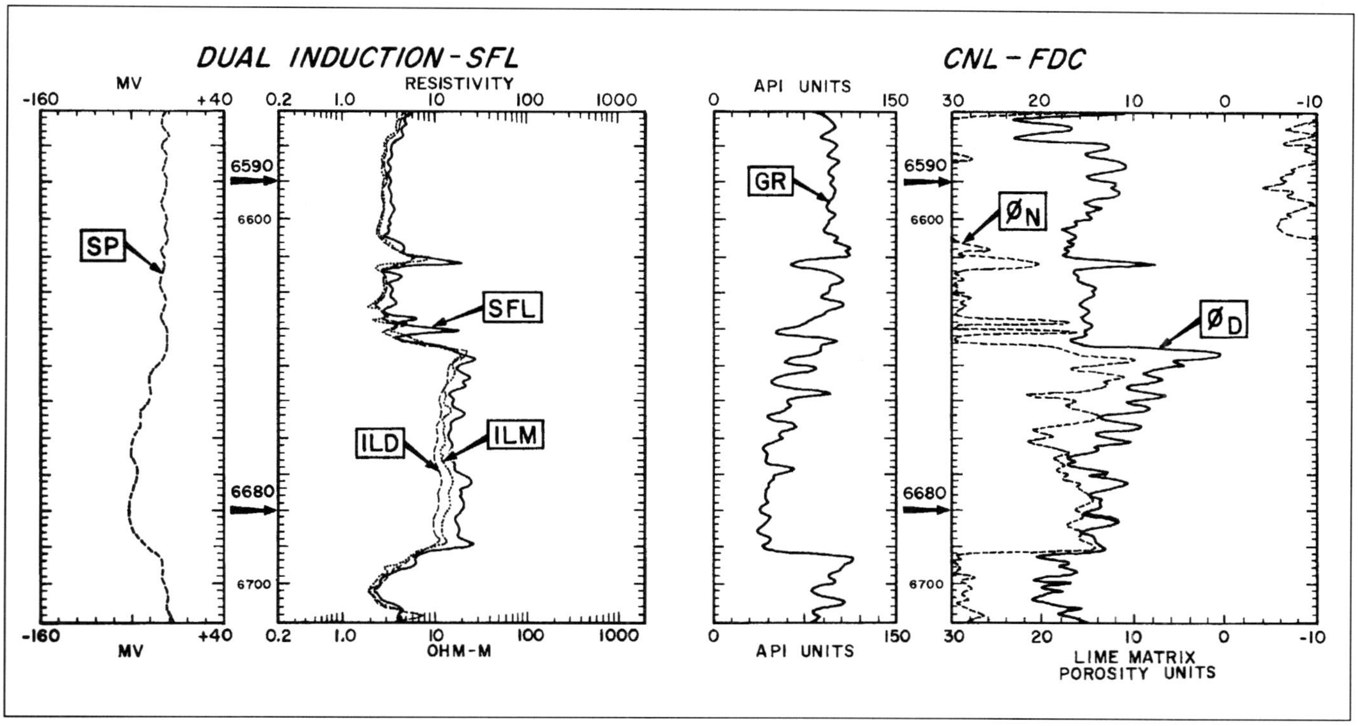

Figure 8—Dual Induction—SFL and gamma ray neutron-lithodensity logs through the Pennsylvanian Red Fork Sandstone, western Oklahoma. In the Red Fork Sandstone at a depth of 6680 feet the following values are recorded:
$\phi_n = 15\%$
$\phi_d = 13\%$
$R_t = 11\Omega - M$
$R_{SFL} = 20\Omega - M$
At a depth of 6590 feet the following shale values are recorded:
$R_{sh} = 3\Omega - M$
$GR_{max} = 98$ API units
$\phi_{nsh} = 33\%$
$\phi_{dsh} = 12\%$

Development of Shaly Sand Analysis

1950s–1980s

The formulas used to correct total water saturation (S_{wt}) to effective water saturation (S_{we}) depend on availability of certain logging suites, on whether or not the distribution of clay is known (i.e. laminated, structural or dispersed), and on access to a computer. The correction formulas are developed around data from different log types and show the evolution brought about as the sophistication of logs increased through the 1950s to 1980s. Some of the formulas require a number of different steps and have so many variables that a manual execution of these steps over any great thickness of strata requires Herculean efforts, hence the need for a computer.

What follows is a classification of formulas by both log type and by date. "New" log types and "new" formulas were added with each decade. However, not all wells today are logged with a complete log suite, and when this is the case, the reader will find it helpful to use formulas that match the available log suite, regardless of the time classification.

Basically, this presentation is designed as an easy reference guide, so that a log can readily be placed within its time frame, or matched by log suite, and then the recommended shaly sand correction techniques applied.

For the purpose of allowing comparison of the differences among the effective water saturation (S_{we}) correction techniques, data from a well penetrating a Pennsylvanian Red Fork shaly sandstone with dispersed clay, located in western Oklahoma, is used in each of the formulas. The reader will find that despite the method used, the results are fairly similar; proving that shaly sand log analysis can be executed even with limited data. All Red Fork shaly sand calculations are taken from a zone at 6680 feet, and logs are selected on the basis of the required data needed for the various shaly sand corrections. The log suite from the Red Fork well consists of the following logs (Figs. 8 and 9):

Dual Induction - SFL (Fig. 8)
Gamma ray neutron-lithodensity (Figs. 8 and 9)
Sonic (Fig. 9)

The various shaly sand log analysis methods, along with the logs needed for their calculation, are broken into an historical classification in Table 4. This table follows the organization of the text as it covers the formulas that correct the log measurements for the presence of shale

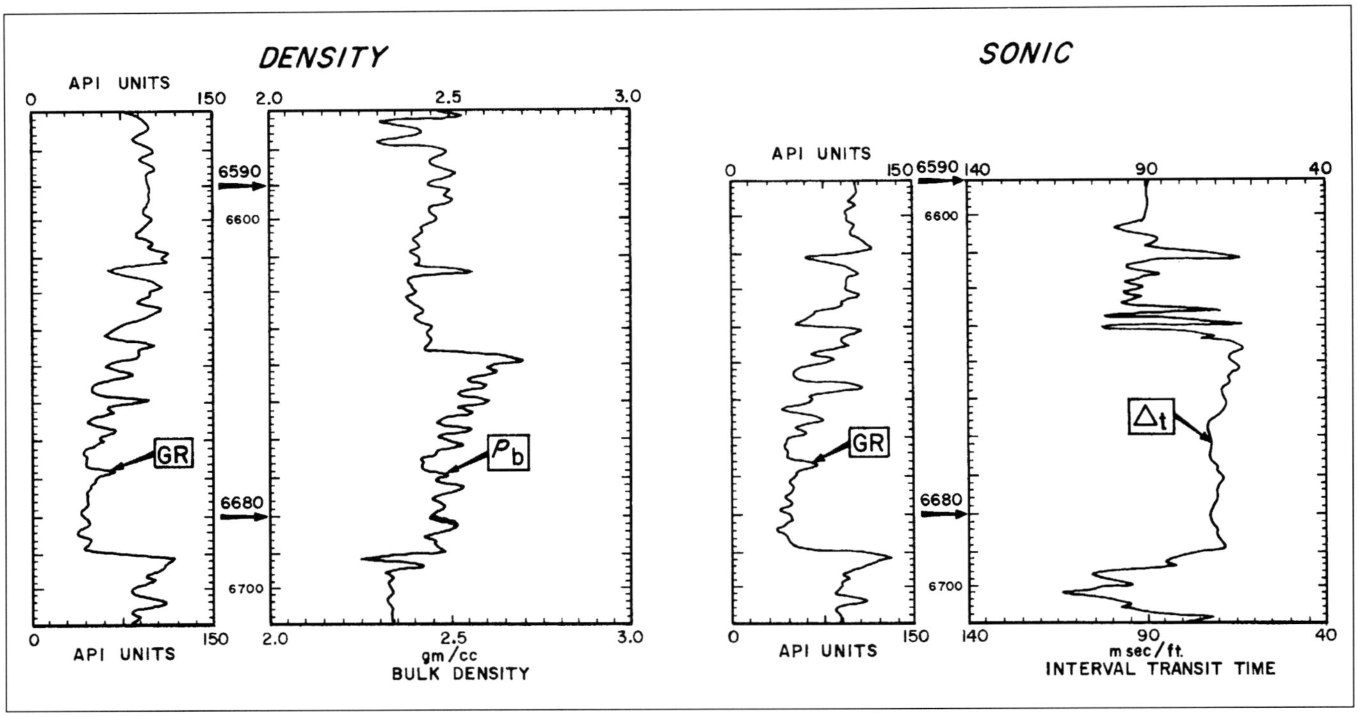

Figure 9—Gamma ray plus sonic and bulk density logs through the Pennsylvanian Red Fork Sandstone, western Oklahoma. In the Red Fork Sandstone at a depth of 6680 feet the following values are recorded:

Δt_{log} = 72 µsec/ft.
ρ_b = 2.48 gm/cc
GR_{log} = 40 API units
At a depth of 6590 feet the following shale values are recorded:
Δt_{sh} = 90 µsec/ft.
ρ_{sh} = 2.50 gm/cc

or clay. Following each time frame are worked examples of the equations.

1950 Logs

In the early 1950s only electric logs were available and porosity had to be determined from these tools (Fig. 8). The formula for calculating resistivity derived porosity (ϕ) is:

$$\phi_t = \left[\frac{a(R_{mf}/R_{xo})}{(S_{xo})^2}\right]^{1/m}$$

where:
ϕ_t = resistivity derived total porosity (uncorrected for clay)
R_{mf} = resistivity of mud filtrate
R_{xo} = shallow resistivity
a = 0.81 consolidated sandstone or 0.62 unconsolidated sand
m = 2.0 consolidated sandstone or 2.15 unconsolidated sand
S_{xo} = water saturation of the flushed zone (see chart below)

	API°	S_{xo}
Gas		60-90
high gravity oil	40-50	90-95
medium gravity oil	20-40	80-90
low gravity oil	10-20	70-80

(Hilchie, 1978)

Effective porosity is then calculated:
$\phi_e = \phi_t \times (1 - V_{cl})$
where:
ϕ_e = resistivity derived effective porosity
ϕ_t = resistivity derived total porosity
V_{cl} = volume of clay

After the resistivity-derived porosity (ϕ_t) and effective porosity (ϕ_e) are calculated, effective water saturation (S_{we}) can be calculated by the No Porosity Log Dual Water and No Porosity Log Automatic Compensation methods.

No Porosity Log Dual Water Method (Elphick, 1988)

First calculate the resistivity of the clay-bound water.

$R_b = R_{sh} \times \phi_t^2$
where:
R_b = resistivity of clay bound water
R_{sh} = resistivity of adjacent shale
ϕ_t = total (resistivity-derived) porosity

Using the Dual Water model, we calculate the resistivity the shaly sand should have if it was 100 percent water saturated (S_w = 1.0). In the Dual Water model the shale or clay portion of the rock has a water resistivity of R_b and the free water (i.e. non-clay bound) has a resistivity of R_w.

$R_o = (R_b \times R_w)/\{[R_w \times V_{cl} + R_b \times (1 - V_{cl})] \times \phi_t^2\}$
assuming a = 1.0 m = 2.0

Table 4. List of Logging Suites and Shaly Sand Methods.

Log Date	Shaly Sand Methods
Early 1950s	
* Gamma ray log	No Porosity Log Dual Water
* Resistivity logs	No Porosity Log Automatic Compensation
* SP log	
Late 1950s	
* Gamma ray log	Automatic Compensation
* Resistivity logs	
* SP log	
* Sonic log	
1960s	
* Gamma ray log	Dispersed Clay
* Resistivity logs	
* SP log	
* Sonic Log	
* Density log	
1970s	
* Gamma ray log	Simandoux
* Resistivity logs	Fertl
* SP log	
Sonic log	
* Neutron-density logs	
1980s	
* Gamma ray log	Dual Water
* Resistivity logs	
* SP log	
Sonic log	
* Neutron-density logs	
Dielectric logs	

* Indicates logs required to calculate specific shaly sand equations.

where:

R_o = wet resistivity of shaly sand
R_b = bound water resistivity
R_w = free water resistivity
V_{cl} = volume of clay
ϕ_t = resistivity-derived total porosity

Now we can calculate effective (clay-corrected) water saturation.

$S_{we} = (R_o/R_t)^{1/2}$

where:

S_{we} = effective (clay-corrected) water saturation
R_o = wet resistivity
R_t = deep resistivity of formation

No Porosity Log Automatic Compensation Method

The No Porosity Log Automatic Compensation method is based on an understanding that the presence of shale or clay causes resistivity (R_t) to be too low and causes resistivity-derived porosity (ϕ_t) to be too high. The diametric nature of these two effects allows them to compensate for each other. The effective water saturation (S_{we}) is calculated by the following formula:

$$S_{we} = \left(\frac{a}{\phi_t^{\,m}} \times \frac{R_w}{R_t}\right)^{1/2}$$

where:

a = 0.81 consolidated sands and 0.62 unconsolidated sands
m = 2.0 consolidated sands and 2.15 unconsolidated sands
R_w = formation water resistivity
R_t = deep resistivity of formation
S_{we} = effective (clay-corrected) water saturation
ϕ_t = resistivity derived porosity uncorrected for clay

Examples of early 1950s methods. Using dual induction log (Fig. 8), first, calculate a volume of clay (V_{cl}) using the gamma ray log at a depth of 6680 feet.

GR_{max} = 98 at 6590 feet (Fig. 8)
GR_{min} = 17 at 6460 feet (Fig. 10)
GR_{log} = 40 at 6680 feet (Fig. 8)

$$I_{GR} = \frac{GR_{log} - GR_{min}}{GR_{max} - GR_{min}}$$

$$I_{GR} = \frac{40 - 17}{98 - 17}$$

$I_{GR} = 0.28$
$V_{cl} = 0.33[2^{(2 \times I_{GR})} - 1.0]$ consolidated sandstones
V_{cl} = 0.16 (16%) at 6680 feet

Next calculate resistivity-derived total porosity (ϕ_t) and resistivity-derived effective porosity (ϕ_e).

$$\phi_t = \left[\frac{0.81\ (R_{mf}/R_{xo})}{S_{xo}}\right]^{1/2}$$

ϕt = 14.0%
$\phi_e = \phi_t \times (1 - V_{cl})$
ϕ_e = 11.8%

V_{cl} = 16% at 6680 feet
R_{mf} = 0.31 @ T_f
R_{xo} = 20 Ω – M (R_{sfl})* at 6680 feet (Fig. 8)
S_{xo} = 0.80 medium gravity oil
a = 0.81 consolidated sandstone
m = 2.0 consolidated sandstone

The clay-bound water resistivity is determined by the formula:

$R_b = R_{sh} \times \phi_t^2$
$R_b = 0.059$

R_{sh} = 3.0 Ω – M at 6590 feet
ϕ_t = 14.0%

Now, we are ready to calculate the wet resistivity (R_o) of the shaly Red Fork sand and the effective water saturation (S_{we}) by the No Porosity Log Dual Water method (Elphick, 1988).

$R_o = (R_b \times R_w)/\{[R_w \times V_{cl} + R_b \times (1 - V_{cl})] \times \phi_t^2\}$
$R_o = 2.6$

R_w = 0.05 at T_f
R_b = 0.059
V_{cl} = 16%
ϕ_t = 14.0%

$S_{we} = (R_o/R_t)^{1/2}$
S_{we} = 48.6%

R_o = 2.6
R_t = 11Ω – M R_{ild}† at 6680 feet

* R_{SFL} was used for R_{xo} because Tornado chart invasion corrections were not made during the early 1950s.

† R_{ild} was used for R_t, because Tornado chart invasion corrections were not made during the early 1950s.

Using the No Porosity Log Automatic Compensation method, the effective water saturation (S_{we}) is calculated as follows:

$$S_{we} = \left(\frac{0.81}{\phi_t^2} \times \frac{R_w}{R_t}\right)^{1/2}$$

$S_{we} = 43.3\%$
$R_w = 0.05$ at T_f
$R_t = 11\Omega$ –M at 6680 feet
$\phi_t = 14.0\%$ at 6680 feet
a = 0.81 consolidated sand
m = 2.00 consolidated sand

Therefore, using only the resistivity log (Fig. 8) the effective (clay-corrected) porosity and water saturations at a depth of 6680 feet in the Red Fork sand are:

$\phi_e = 11.8\%$
$S_{we} = 48.6\%$ No Porosity Log Dual Water Method (Elphick, 1988)
$S_{we} = 43.3\%$ No Porosity Log Automatic Compensation Method

In the late 1950s, a sonic log was added to the log suite and so both the electric log (Fig. 8) and the sonic log (Fig. 9) could be used for shale corrections. The shaly sand analysis technique, based on the electric and sonic log, is called the Automatic Compensation method. This method builds on the knowledge that shale or clay causes the resistivity log to read too low a value for resistivity (R_t) and causes the sonic log to read too high a porosity (ϕ_s). The low R_t, high ϕ_s readings compensate for each other. The electric log/sonic log shaly sand correction formulas, called the Automatic Compensation method, follow:

$$S_{we} = \left(\frac{0.81}{\phi_s^2} \times \frac{R_w}{R_t}\right)^{1/2}$$

$$\phi_s = \frac{\Delta t_{log} - \Delta t_{ma}}{\Delta t_f - \Delta t_{ma}} \times \frac{100}{\Delta t_{sh}}$$

$$\phi_e = \left(\frac{\Delta t_{log} - \Delta t_{ma}}{\Delta t_f - \Delta t_{log}} \times \frac{100}{\Delta t_{sh}}\right) - V_{cl}\left(\frac{\Delta t_{sh} - \Delta t_{ma}}{\Delta t_f - \Delta t_{ma}}\right)$$

where:
ϕ_s = sonic porosity uncorrected for clay (i.e. total porosity)
R_w = formation water resistivity
R_t = formation deep resistivity
ϕ_e = effective porosity from sonic log (corrected for clay)
S_{we} = effective water saturation
Δt_{log} = interval transit time shaly sand
Δt_f = interval transit time fluid
Δt_{ma} = interval transit time matrix
Δt_{sh} = interval transit time shale
V_{cl} = volume of clay

$\frac{100}{\Delta t_{sh}}$ = a compaction correction only valid for unconsolidated sands

Example of Late 1950s Method. The dual induction log (Fig. 8) and the sonic log (Fig. 9) are used to calculate total sonic porosity (ϕ_s), effective sonic porosity (ϕ_e), and effective water saturation (S_{we}), at a depth of 6680 feet, by the Automatic Compensation method.

$$\phi_s = \frac{\Delta t_{log} - \Delta t_{ma}}{\Delta t_f - \Delta t_{ma}}$$

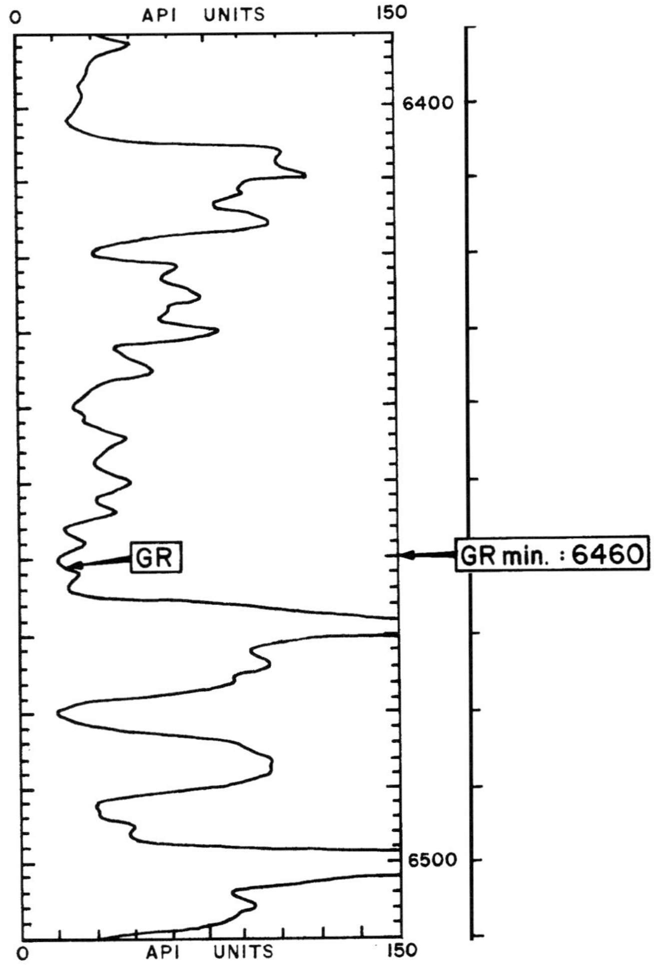

Figure 10—Gamma ray log from the same well as figures 8 and 9 but logged over a clean sand uphole from the Red Fork Sandstone. A gamma ray minimum value of 17 API units is recorded at a depth of 6460 feet.

$\phi_s = \phi_t = 15.2\%$

$$\phi_e = \left(\frac{\Delta t_{log} - \Delta t_{ma}}{\Delta t_f - \Delta t_{ma}}\right) - V_{cl}\left(\frac{\Delta t_{sh} - \Delta t_{ma}}{\Delta t_f - \Delta t_{ma}}\right)$$

$\phi_e = 10.7\%$
$\Delta t_{log} = 72$ μsec/ft. at 6680 feet
$\Delta t_{ma} = 51$ μsec/ft. sandstone
$\Delta t_f = 189$ μsec/ft. fresh water mud
$V_{cl} = 16\%$ at 6680 feet already calculated
$\Delta t_{sh} = 90$ μsec/ft. at 6590 feet

The effective (clay-corrected) water saturation (S_{we}) is then calculated.

$$S_{we} = \left(\frac{0.81}{\phi_s^2} \times \frac{R_w}{R_t}\right)^{1/2}$$

$S_{we} = 39.9\%$
$\phi_s = 15.2\%$ at 6680 feet
$R_t = 11\Omega$ – M at 6680 feet
$R_w = 0.05$

Using the dual induction log (Fig. 8) and the sonic log (Fig. 9) the effective porosity (ϕ_e) and effective water saturation (S_{we}), calculated at a depth of 6680 feet in the Red Fork sand, are:

$S_{we} = 39.9\%$
$\phi_e = 10.7\%$

1960 Logs

In the 1960s, density logs were added to resistivity, and sonic logs (Figs. 8 and 9), and the shaly sand technique for using them was called the Dispersed Clay method (Alger et al., 1963). The sonic log detects the dispersed clay in the pore water as a slurry and gives porosity a value equal to the sum of the volumetric fraction (total porosity ϕ_t). The density log detects only water-filled porosity (Dewan, 1983). The fraction of the clean-sand intergranular pore space occupied by clay is called q. Where:

$q = (\phi_s - \phi_d)/\phi_s$

ϕ_s = sonic porosity uncorrected for clay

ϕ_d = density porosity uncorrected for clay

Effective water saturation is:

$$S_{we} = \left[\sqrt{\frac{0.8}{\phi_s^2} \times \frac{R_w}{R_t} + \left(\frac{q}{2}\right)^2} - \frac{q}{2}\right] /(1-q)$$

(Alger et al., 1963)

where:

S_{we} = effective (clay-corrected) water saturation

ϕ_s = sonic porosity uncorrected for clay

R_w = formation water resistivity

$q = (\phi_s - \phi_d)/\phi_s$

R_t = deep formation resistivity

An important aspect of the Dispersed Clay method is that the equation does not require a value for shale resistivity (R_{sh}) or volume of clay (V_{cl}), because the shaly factor (q) is determined within the shaly sand. Effective porosity is:

$$\phi_e = \left(\frac{\rho_{ma} - \rho_b}{\rho_{ma} - \rho_f}\right) - V_{cl}\left(\frac{\rho_{ma} - \rho_{sh}}{\rho_{ma} - \rho_f}\right)$$

where:

ϕ_e = effective porosity corrected for clay

ρ_{ma} = matrix density

ρ_b = bulk density from shaly sand

ρ_f = fluid density

ρ_{sh} = shale density

V_{cl} = volume of clay

Example of 1960s Method. The dual induction log (Fig. 8), sonic log (Fig. 9), and density log (Fig. 9) are used to calculate effective porosity (ϕ_e) and effective water saturation (S_{we}) by the Dispersed Clay method at a depth of 6680 feet, in the Red Fork example.

First sonic and density porosity and a value for q are calculated.

$\phi_s = \phi_t = 15.2\%$ already calculated in Automatic Compensation method

$$\phi_d = \frac{\rho_{ma} - \rho_b}{\rho_{ma} - \rho_f}$$

$\phi_d = 11.9\%$

$q = (\phi_s - \phi_d)/\phi_s$

$q = 0.22$

ρ_{ma} = 2.68 gm/cc sandstone matrix

ρ_b = 2.48 gm/cc at 6680 feet

ρ_f = 1.0 gm/cc fresh water mud

Next effective porosity (ϕ_e) from the density log is calculated.

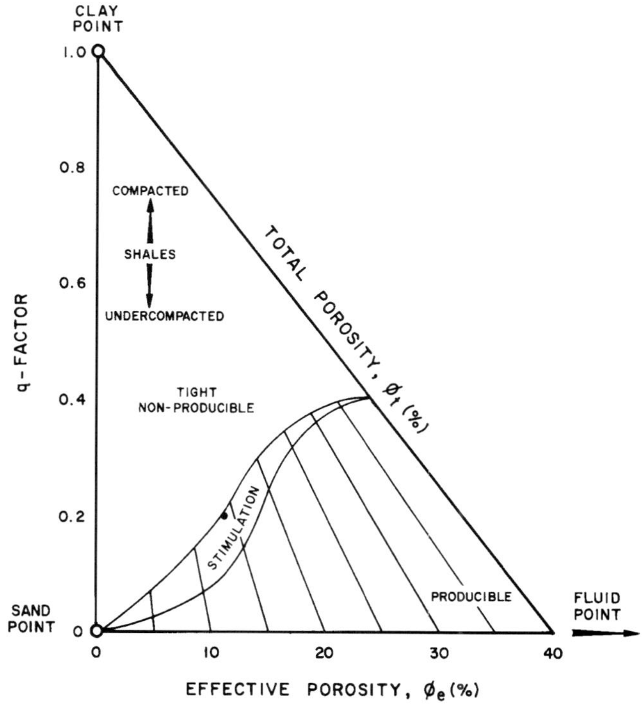

Figure 11—Shaly Sand Producibility chart. The data plotted in Figure 11 are from the Red Fork example at a depth of 6680 feet. The data from the Red Fork Sandstone plot on the chart just inside the producible area, indicating that the Red Fork Sandstone should be a marginal producer unless it is stimulated.
ϕ_e = effective (clay-corrected) porosity
q = intergranular pore space occupied by clay
$q = (\phi_s - \phi_d)/\phi_s$
The Shaly Sand Producibility Chart is based on field data from U.S. Gulf Coast, New Mexico, Colorado, and Wyoming (Dresser Atlas, 1979), and is designed for use in shaly reservoirs with dispersed clay.

$$\phi_e = \left(\frac{\rho_{ma} - \rho_b}{\rho_{ma} - \rho_f}\right) - V_{cl}\left(\frac{\rho_{ma} - \rho_{sh}}{\rho_{ma} - \rho_f}\right)$$

$\phi_e = 10.2\%$

V_{cl} = 16% already calculated at 6680 feet

ρ_{ma} = 2.68 gm/cc sandstone

ρ_{sh} = 2.50 gm/cc at 6590 feet

ρ_b = 2.48 gm/cc at 6680 feet

Last, effective water saturation (S_{we}) is found by the Dispersed Clay method.

$$S_{we} = \left[\sqrt{\frac{0.8}{\phi_s^2} \times \frac{R_w}{R_t} + \left(\frac{q}{2}\right)^2} - \frac{q}{2}\right] /(1-q)$$

$S_{we} = 38.7\%$

$q = 0.22$ at 6680 feet

$\phi_s = 15.2\%$ at 6680 feet

$R_w = 0.05$

$R_t = 11\Omega - M$ at 6680 feet

Using the dual induction log (Fig. 8), sonic log (Fig. 9) and density log (Fig. 9), the effective porosity (ϕ_e) and effective water saturation (S_{we}) at a depth of 6680 feet in Red Fork sand are:

$S_{we} = 38.7\%$

$\phi_e = 10.2\%$

The Dispersed Clay method may have lower reliability in gas sands, because density porosity may be greater than sonic porosity (remember that the q equation assumes $\phi_s > \phi_d$).

The value for q and effective porosity (ϕ_e) can be cross-plotted on a Shaly Sand Producibility chart (Fig. 11), to determine if the volume of clay is high enough to drastically reduce producibility. A well will not produce if the data plot in the nonproducible part of the cross plot (Fig. 11). In addition to calculating q using the sonic (ϕ_s) and density (ϕ_d) porosities, q can also be calculated by the following formulas:

$q = (\phi_t - \phi_e)/\phi_t$

$q = aV_{cl}/\phi_s$

$q = aV_{cl}/(\phi_d + aV_{cl})$

where:

V_{cl} = volume of clay
ϕ_t = total porosity (uncorrected for clay)
ϕ_e = effective porosity (corrected for clay)
ϕ_d = density porosity
ϕ_s = sonic porosity
a = 0.35 consolidated sandstone and 0.25 unconsolidated sand

Any of the values listed for q can also be cross-plotted with effective porosity (ϕ_e) to determine producibility (Fig. 11). The ability to determine q by various methods, means that ϕ_e and q Shaly Sand Producibility cross-plots can be used with any logging suite.

1970 Logs

In the 1970s, the typical log suite included electric (Fig. 8) and combination neutron-density logs (Fig. 8). Two of the commonly used shaly sand equations are the Simandoux (1963) and Fertl (1975) methods.

$$S_{we} = \frac{C \times R_w}{\phi_e^2}\left[\sqrt{\frac{5\phi_e^2}{R_w \times R_t} + \left(\frac{V_{cl}}{R_{sh}}\right)^2} - \frac{V_{cl}}{R_{sh}}\right]$$

(Simandoux, 1963)

where:

S_{we} = effective (clay-corrected) water saturation
ϕ_e = effective porosity corrected for clay
R_{sh} = resistivity of adjacent shale
R_w = formation water resistivity
R_t = deep formation resistivity
V_{cl} = volume of clay
C = 0.40 for sandstones and 0.45 for carbonates

The neutron and density porosities are corrected for clay by the following formulas:

$\phi_{dc} = \phi_d - (V_{cl} \times \phi_{dsh})$

$\phi_{nc} = \phi_n - (V_{cl} \times \phi_{nsh})$

$\phi_e = (\phi_{dc} + \phi_{nc})/2$ for oil zones

$\phi_e = [(\phi_{dc}^2 + \phi_{nc}^2)/2]^{1/2}$ for gas zones

where:

ϕ_d = density porosity
ϕ_n = neutron porosity
ϕ_{dc} = density porosity corrected for clay
ϕ_{nc} = neutron porosity corrected for clay
ϕ_e = effective porosity corrected for clay
V_{cl} = volume of clay
ϕ_{dsh} = density porosity of shale
ϕ_{nsh} = neutron porosity of shale

$$S_{we} = \frac{1}{\phi_d}\left[\sqrt{\frac{R_w}{R_t} + \left(\frac{a \times V_{cl}}{2}\right)^2} - \frac{a \times V_{cl}}{2}\right]$$

(Fertl, 1975)

where:

S_{we} = effective (clay-corrected) water saturation
ϕ_d = density porosity uncorrected for clay
R_w = formation water resistivity
R_t = deep formation resistivity
V_{cl} = volume of clay
a = 0.25 Gulf Coast and 0.35 Rocky Mountains
0.81 R_w/R_t used in the Gulf Coast

The effective porosity (ϕ_e) is calculated by the same formulas used with the Simandoux equation.

Examples of 1970s Methods. Using the dual induction log (Fig. 8) and the combination neutron-density log (Fig. 8), we calculate effective porosity (ϕ_e) and effective water saturation (S_{we}) at a depth of 6680 feet by both the Simandoux and Fertl methods. First, we calculate the density porosity (ϕ_d) and the effective neutron-density porosity (ϕ_e).

$$\phi_d = \frac{\rho_{ma} - \rho_b}{\rho_{ma} - \rho_f}$$

ϕ_d = 11.9%
ρ_{ma} = 2.68 gm/cc sandstone
ρ_f = 1.0 gm/cc fresh water mud
ρ_b = 2.48 gm/cc at 6680 feet

$\phi_{nc} = \phi_n - (V_{cl} \times \phi_{nsh})$

ϕ_{nc} = 9.7%

$\phi_{dc} = \phi_d - (V_{cl} \times \phi_{dsh})$

ϕ_{dc} = 11.1%

$\phi_e = (\phi_{nc} + \phi_{dc})/2$ oil zone

ϕ_e = 10.4%
V_{cl} = 16% at 6680 feet already calculated
ϕ_{nsh} = 33% at 6590 feet
ϕ_{dsh} = 12% at 6590 feet
ϕ_n = 15% at 6680 feet
ϕ_d = 13% at 6680 feet

Simandoux Method:

$$S_{we} = \frac{C \times R_w}{\phi_e^2}\left[\sqrt{\frac{5\phi_e^2}{R_w \times R_t} + \left(\frac{V_{cl}}{R_{sh}}\right)^2} - \frac{V_{cl}}{R_{sh}}\right]$$

S_{we} = 48.9%
ϕ_e = 10.4% at 6680 feet
C = 0.4 sandstone
R_{sh} = 3.0Ω – M at 6590 feet
R_w = 0.05
R_t* = 11Ω – M at 6680 feet
V_{cl} = 16% at 6680 feet

Fertl Method:

$$S_{we} = \frac{1}{\phi_d}\left[\sqrt{\frac{R_w}{R_t} + \left(\frac{a \times V_{cl}}{2}\right)^2} - \frac{a \times V_{cl}}{2}\right]$$

S_{we} = 37.8%
ϕ_d = 11.9% at 6680 feet
R_w = 0.05
R_t = 11Ω – M at 6680 feet

*R_{ild} was used for R_t, because the invasion correction is very small (i.e., R_{ild} = 11Ω – M/R_t = 10.2Ω – M).

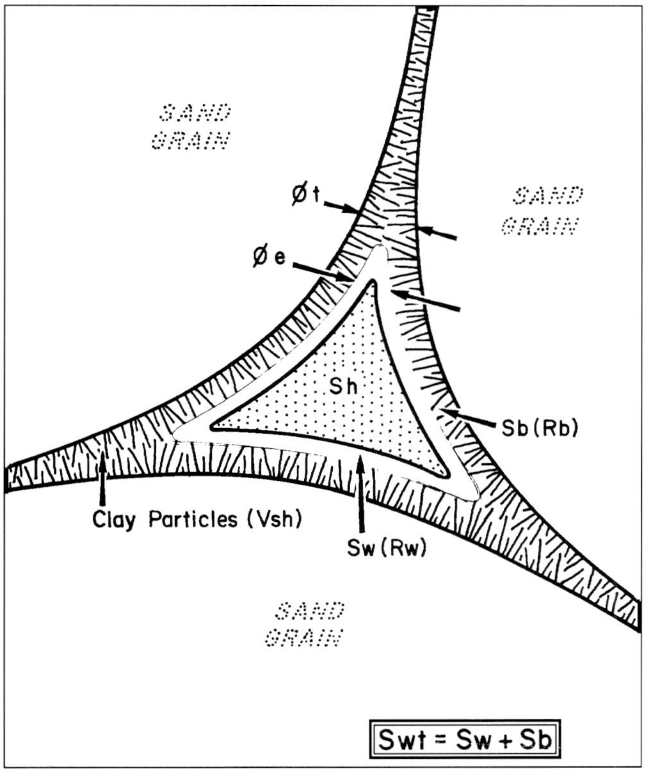

Figure 12—Partitioning of a shaly sand into clay bound water (S_b) and free water (S_w) in the Dual Water model. The free water (S_w) may or may not be free to move, because adhesion and capillary forces can still bind non-clay bound water. The presence of clay-bound water (S_b) in the pores reduces the effective size of the pores, which, in turn, also reduces permeability.

a = 0.35 consolidated
V_{cl} = 16% at 6680 feet

Calculating effective porosity (ϕ_e) from the neutron-density log (Fig. 8) and effective water saturations (S_{we}) by Simandoux and Fertl methods for the Red Fork sand at a depth of 6680 feet gives the following results:

S_{we} = 48.9% Simandoux
S_{we} = 37.8% Fertl
ϕ_e = 10.4%

The Fertl technique has two advantages over the Simandoux technique: l) it is easier to calculate, and 2) it does not need a value for resistivity of the adjacent shale, which can be important in areas of consolidated sands and shales. As noted earlier, often the resistivity of the adjacent shale is higher than the shaly sandstone.

1980 Logs and Computers

In the 1980s, because of the increasing use of computers, several complex shaly sand equations became popular. Also, a transformation in shaly sand analysis was initiated. The basis of this transformation was an attempt to use cation exchange capacity (CEC) instead of volume of clay (V_{cl}) in shaly sand equations. Although the cation exchange capacity is a better measure of clay's impact on logs than is V_{cl}, it must be laboratory-determined and depends on core data. An example of a CEC based model is the Waxman-Smit model (Waxman and Smit, 1968). Major objections to the Waxman-Smit model stem from the fact that it is difficult to use without laboratory data (Hilchie, 1982), and that it predicts that water sands of constant R_w, but increasing shaliness, have decreasing effective water resistivities (R_w). Clavier et al. (1977) have found a good deal of evidence to the contrary. The difficulties incurred by the Waxman-Smit model have led us to select a model for a 1980s log analysis method called the Dual Water method. It circumvents the flaws in the Waxman-Smit model, and is indirectly based on cation exchange capacity (Schlumberger, 1987).

The Dual Water model (Clavier, et. al., 1977) has pore water partitioned into bound water (S_b) and free water (S_w), each of which contributes to the resistivity (conductivity) of the shaly sand (Fig. 12). Also, the bound water (S_b) and the free water (S_w) have their own formation water resistivities (R_b and R_w). The following is a summary of the Dual Water formulas:

1. Calculate volume of clay (V_{cl}): see previous examples.

2. Correct neutron and density porosities for clay.

$\phi_{dc} = \phi_d - (V_{cl} \times \phi_{dsh})$
$\phi_{nc} = \phi_n - (V_{cl} \times \phi_{nsh})$

where:
V_{cl} = volume of clay
ϕ_d = density porosity
ϕ_n = neutron porosity
ϕ_{nc} = neutron porosity corrected for clay
ϕ_{dc} = density porosity corrected for clay
ϕ_{dsh} = density porosity of shale
ϕ_{nsh} = neutron porosity of shale

3. Calculate effective porosity (ϕ_e).

$\phi_e = (\phi_{nc} + \phi_{dc})/2$ oil zone
$\phi_e = [(\phi_{nc}^2 + \phi_{dc}^2)/2]^{1/2}$ gas zone

where:
ϕ_e = effective porosity corrected for clay
ϕ_{nc} = neutron porosity corrected for clay
ϕ_{dc} = density porosity corrected for clay

4. Calculate total porosity of adjacent shale (ϕ_{tsh}).

$\phi_{tsh} = \delta\phi_{dsh} + (1 - \delta)\,\phi_{nsh}$

where:
δ = 0.5 to 1.0
ϕ_{dsh} = density porosity of shale
ϕ_{nsh} = neutron porosity of shale

5. Calculate total porosity (ϕ_t) and bound water saturation (S_b).

$\phi_t = \phi_e + V_{cl} \times \phi_{tsh}$
$S_b = V_{cl} \times \phi_{tsh}/\phi_t$

where:
S_b = clay-bound water saturation
ϕ_t = total porosity
ϕ_e = effective porosity
V_{cl} = volume of clay
ϕ_{tsh} = total porosity of adjacent shale

6. Calculate bound water resistivity (R_b) from adjacent shale.

$R_b = R_{sh} \times \phi_{tsh}^2$

where:
R_b = bound water resistivity
R_{sh} = resistivity of adjacent shale

Table 5. Summary of Effective Porosities (ϕ_e) and Effective Water Saturations (S_{we}) for the Pennsylvanian Red Fork Sandstone from a Depth of 6680 Feet Using Different Shaly Sand Equations and Different Porosity Logs.

	ϕ_e	S_{we}	S_{we}*
NO POROSITY LOG (DUAL WATER)	11.8% (resistivity porosity)	48.6%	43.7%
NO POROSITY LOG (AUTOMATIC COMPENSATION)	11.8% (resistivity porosity)	43.3%	
AUTOMATIC COMPENSATION	10.7% (sonic log)	39.9%	
DISPERSED CLAY	10.2% (density log)	38.7%	
SIMANDOUX	10.4% (neutron-density log)	48.9%	38.5%
FERTL	10.4% (neutron-density log)	37.8%	
DUAL WATER	10.4% (neutron-density log)	44.1%	33.5%
	ϕ_t	S_{wt}	
NO SHALE CORRECTION	11.9% (density porosity)	51.0%	

S_{we}*—assuming resistivity of dispersed clay (R_{cl}) = 0.4 R_{sh} (Fertl and Hammack, 1971).
R_{cl} = 0.4 x 3 Ω - M
R_{cl} = 1.2 Ω - M at 6590 feet

ϕ_{tsh} = total porosity of adjacent shale

7. Calculate apparent water resistivity (R_{wa}) in the shaly sand.

$$R_{wa} = R_t \times \phi_t^2$$

where:
R_{wa} = apparent formation water resistivity
R_t = deep formation resistivity
ϕ_t = total porosity

8. Calculate total water saturation (S_{wt}) corrected for clay.

$$S_{wt} = b + \sqrt{b^2 + (R_w/R_{wa})}$$

where:
S_{wt} = total water saturation corrected for clay
$b = [S_b\,(1 - R_w/R_b)]/2$
R_w = formation water resistivity
R_{wa} = apparent formation water resistivity

9. Calculate effective water saturation (S_{we}) of the shaly sand.

$$S_{we} = (S_{wt} - S_b)/(1 - S_b)$$

where:
S_{we} = effective water saturation
S_{wt} = total water saturation
S_b = clay bound water saturation

Example of 1980s Method. Using the dual induction log (Fig. 8) and the combination neutron-density log (Fig. 8), we can calculate effective neutron-density porosity (ϕ_e) and effective water saturation (S_{we}) from the Dual Water equations, at a depth of 6680 feet. We begin by determining volume of clay and calculating effective (ϕ_e) neutron-density porosity (steps 1–3 of the Dual Water method).

First, we determine volume of clay (V_{cl}), then we correct neutron and density porosities for clay, and finally, we calculate effective porosity (ϕ_e).

1. Calculate volume of clay (V_{cl}).
V_{cl} = 16% previously determined
2. Correct neutron and density porosities for clay.
$\phi_{nc} = \phi_n - (V_{cl} \times \phi_{nsh})$
$\phi_{dc} = \phi_d - (V_{cl} \times \phi_{dsh})$
ϕ_{nc} = 9. 7%
ϕ_{dc} = 11.1%
ϕ_n = 15% at 6680 feet
ϕ_d = 13% at 6680 feet
V_{cl} = 16% at 6680 feet already determined
ϕ_{nsh} = 33% at 6590 feet
ϕ_{dsh} = 12% at 6590 feet
3. Calculate effective porosity (ϕ_e).
$\phi_e = (\phi_{nc} + \phi_{dc})\,/2$ oil zone
ϕ_e = 10.4%
4. Calculate total porosity of adjacent shale (ϕ_{tsh}).
$\phi_{tsh} = \delta\phi_{dsh} + (1 - \delta)\phi_{nsh}$
ϕ_{tsh} = 18.3%
δ = 0.7
ϕ_{dsh} = 12% at 6590 feet
ϕ_{nsh} = 33% at 6590 feet

5. Calculate total porosity (ϕ_t) and bound water saturation (S_b).

$\phi_t = \phi_e + V_{cl} \times \phi_{tsh}$

$\phi_t = 13.3\%$

$\phi_e = 10.4\%$ at 6680 feet already determined

$V_{cl} = 16\%$ at 6680 feet already determined

$\phi_{tsh} = 18.3\%$ at 6590 feet already determined

$S_b = V_{cl} \times \phi_{tsh} / \phi_t$

$S_b = 22\%$

$\phi_t = 13.3\%$ at 6680 feet

$V_{cl} = 16\%$ at 6680 feet

$\phi_{tsh} = 18.3\%$ at 6590 feet

6. Calculate bound water resistivity (R_b) from adjacent shale.

$R_b = R_{sh} \times \phi_{tsh}^2$

$R_b = 0.10$

$R_{sh} = 3.0\Omega - M$ at 6590 feet

$\phi_{tsh} = 18.3\%$ at 6590 feet

7. Calculate apparent water resistivity (R_{wa}) in the shaly sand.

$R_{wa} = R_t \times \phi_t^2$

$R_{wa} = 0.195$

$\phi_t = 13.3\%$ at 6680 feet

$R_t = 11\Omega - M$ at 6680 feet

$$S_{wt} = b + \sqrt{b^2 + (R_w/R_{wa})}$$

8. Calculate total water saturation (S_{wt}) corrected for clay.

$S_{wt} = 56.4\%$

$R_w = 0.05$

$R_{wa} = 0.195$ at 6680 feet

$b = [S_b\ (1 - R_w/R_b)]/2$

$S_b = 22\%$ at 6680 feet

$R_b = 0.10$ at 6590 feet

$b = 0.055$

9. Calculate the effective water saturation (S_{we}) of the shaly sand.

$S_{we} = (S_{wt} - S_b)/(1 - S_b)$

$S_{we} = 44.1\%$

$S_{wt} = 56.4\%$ at 6680 feet

$S_b = 22\%$ at 6680 feet

When we use the Dual Water equations to calculate effective neutron-density porosity (ϕ_e) and effective water saturation (Swe) for the Red Fork sand, at a depth of 6680 feet, we find the following:

$S_{we} = 44.1\%$

$\phi_e = 10.4\%$

A summary of the effective porosities (ϕ_e) and effective water saturations (S_{we}), determined by each of the shaly sand methods for the Pennsylvanian Red Fork sand at a depth of 6680 feet (Fig. 8), is presented in Table 5. It is important to note in Table 5 that the equations requiring a value for shale resistivity (i.e. No Porosity Log Dual Water, Simandoux, and Dual Water) have the highest effective water saturations. The reason these equations result in high effective water saturations is that the Red Fork Sandstone example presented here has dispersed clay. The shale resistivity (R_{sh}) from the allogenic adjacent shale used in these equations has a higher resistivity than the resistivity of the dispersed authigenic clay (R_{cl}) in the shaly Red Fork Sandstone. When we calculate the effective water saturations, using the equations that require a value for R_{sh}, Red Fork effective water saturations are too high (Table 5).

Remember that shale resistivity (R_{sh}) is only equal to the resistivity of the clay in the shaly sand when the clay in the sandstone is depositional (allogenic). It is suggested that whenever effective water saturations, calculated from equations that require R_{sh}, are much higher than the effective water saturations calculated from equations that do not require a value for R_{sh}, then dispersed authigenic clay is probably present.

Fertl and Hammack (1971) propose that the resistivity of the dispersed authigenic clay can be approximated by:

$R_{cl} = 0.4 \times R_{sh}$

where:

R_{cl} = dispersed authigenic clay resistivity

R_{sh} = adjacent allogenic shale resistivity

Using the Fertl and Hammack (1971) formula for dispersed clay resistivity for the Red Fork Sandstone, the effective water saturations for the No Porosity Log Dual Water, Simandoux, and Dual Water methods are lower (Table 5).

Shaly Sand Log Analysis Guidelines

The reader will find the following guidelines helpful when selecting a shaly sand log analysis method. The author applies the following rules:

1. When there is no porosity log and shale distribution is laminated, use the No Porosity Log Dual Water method (Elphick, 1988); if the clay is either dispersed or if you do not know the shale or clay distribution, use the No Porosity Log Automatic Compensation method.
2. If the only porosity log is a sonic log, use the Automatic Compensation method for either dispersed or laminated shale. This method, however, works best in medium to high-porosity sandstone with dispersed clay (Dewan, 1983).
3. When the clay is dispersed or you do not know the clay or shale distribution, use either the Dispersed Clay or Fertl methods.
4. If the shale is laminated, use the Simandoux or Dual Water methods.
5. When the clay is dispersed and the resistivity of the dispersed clay is known to be the same as the adjacent shale (bioturbated or turbidite shaly sands), use Simandoux or Dual Water methods.
6. If the resistivity of the dispersed clay in the shaly sand is known or you want to assume that the resistivity of the dispersed clay (R_{cl}) is equal to 0.4 times (Fertl and Hammack, 1971) the resistivity of the adjacent shale (R_{sh}), use Simandoux or Dual Water methods.
7. In areas where formation water resistivities are high ($R_w > 0.1\ \Omega - M$), water saturations calculated by the Fertl (1975) and Dispersed Clay methods should be used with caution. The reason for caution is that it is assumed in both equations that $R_{sh} >> R_w$; however, in shaly sands with high formation water resistivities, this assumption may not be true. If $R_{sh} \cong R_w$ use the Dual Water method.
8. For additional shaly sand and V_{cl} formulas along with guidelines for their use (see: Appendix I and II).

The reason the author prefers the Dispersed Clay or

Fertl methods when the clay is dispersed or when the clay or shale distribution is unknown is that these equations do not require a value for shale resistivity. This absence of need for a shale resistivity value is important because the resistivity of dispersed clay in a reservoir very often is significantly different from the resistivity of the adjacent shale. Dispersed clays are predominantly authigenic and vary in resistivity from adjacent shale beds. This difference in resistivities between dispersed clay in the reservoir and the resistivity of the adjacent shale is a particularly acute problem in shaly sandstones where the resistivity of the adjacent shale is greater than the resistivity of the shaly sands. Shaly sand water saturations, calculated with shaly sand equations requiring shale resistivity (R_{sh}), will have a higher water saturation than by the standard Archie equation technique, whenever the resistivity of the adjacent shale is greater than the resistivity of the shaly sand. Patchett and Herrick (1982) report that, in general, a V_{sh}/R_{sh} term in a shaly sand equation is appropriate only in laminated shaly sands. Therefore, whenever it has been established that the shale distribution is laminated, the author uses either Simandoux or Dual Water methods. The reason that shaly sand equations that require a value for R_{sh} need to be restricted to laminated shaly sands (unless R_{cl} is known) is that the clays in the shaly laminated sand and the adjacent shale are both depositional in origin (allogenic).

Even with all the equations we have discussed, shaly sand analysis is still part science—part art. The most critical parameter V_{cl} is often difficult to determine accurately, and the determination of total shale porosity (ϕ_{tsh}) is still questionable (Dewan, 1983). Also, the problem of determining the resistivity of the dispersed clay (R_{cl}) is still not solved. There is no substitute for experience in an area.

CHAPTER 4

BULK VOLUME WATER AND POROSITY MAPPING IN SHALY SANDSTONES

General

Adjusting log data by a series of formulas, as we have done in the preceding chapter, resolves only one of the issues concerning shaly sands. Another problem is establishing a way to predict, on a broader scale than just the well bore, where shale is degrading a reservoir the most. Changes in reservoir quality are best reflected by identifying special parameters that can be used to create maps. Two of the most significant mapping parameters are porosity and bulk volume water; both of these can be related to log-derived permeability.

Bulk Volume Water, Porosity and Permeability

Bulk volume water is the product of water saturation (S_w) and porosity (ϕ). When a homogeneous reservoir is at irreducible water saturation (i.e. will not produce water), the bulk volume water values are constant. Figure 13 is a cross plot of porosity versus water saturation. Data from a homogeneous reservoir at irreducible water saturation, if plotted on Figure 13, plot either along or parallel to one of the hyperbolic lines representing lines of equal bulk volume water. One use of bulk volume water, then, is to determine when a reservoir is at irreducible water saturation.

Another use of bulk volume water (BVW) is the determination of log-derived permeability (Fig. 13). Log-derived permeability can be found by cross-plotting porosity and water saturation on a log-derived permeability chart (Fig. 13). The chart demonstrates that permeability is a function of both porosity and BVW.

Fertl and Vercellino (1978) note a relationship between grain size and bulk volume water. They find that as grain size of a sand reservoir decreases, there is a concomitant increase in bulk volume water. This relationship holds true because coarser grained sands have larger pores, meaning higher permeability, than more finely grained sands; larger pores have lower irreducible water saturation, meaning they have lower bulk volume water values.

A causative factor that dictates changes in irreducible water saturation and permeability is the presence or absence of clays (Dewan, 1983). As clay content increases, irreducible water saturation and bulk volume water (BVW) increase, but permeability decreases (see: Fig. 12). All of these have a negative effect on reservoir producibility.

Because of the interrelationship between porosity and bulk volume water (Fig. 13) and their control on permeability, both must be considered in order to adequately map the more permeable portions of a reservoir.

Just how much of an effect shale or clay can have on a reservoir, at what point its presence condemns a reservoir for commercial hydrocarbon production, and what mapping strategies predict better reservoir are all graphically described by a detailed Red Fork sand log analysis project.

Bulk Volume Water (BVW) and Porosity Mapping of Shaly Red Fork Sand in Oklahoma

Introduction

The Pennsylvanian Red Fork reservoirs of the Anadarko basin in Oklahoma are predominantly fine-grained, shaly sands that often have high initial flow potential in excess of 3.0 MMCFGPD. Some of the Red Fork high-flow reservoirs have large reserves (>20.0 BCF), but others do not.

Table 6 is a tabulation of production data and net porosity thicknesses of three upper Red Fork wells from the Clinton Field (T12N-R15 and 16W) in Custer County, Oklahoma (Fig. 14). This table illustrates that Red Fork wells can have very similar initial flow rates (compare: Anson #1-18 Murphy and the Conoco #l-A Snider; Table 6), and still have very different ultimate reserves. Table 6 also illustrates that net porosity thicknesses can be very inaccurate for predicting reserves. Note on Table 6 that all three of the Red Fork wells have 30 feet of porosity ≥ 10 percent, but their recoverable reserves vary considerably.

The data in Table 6 illustrate that mapping only net porosity values is not sufficient to accurately define where the best Red Fork reservoirs can be found along a Red Fork trend. What is needed is a more accurate mapping parameter of Red Fork permeability, so that the areas of better long-range, cumulative production can be delineated.

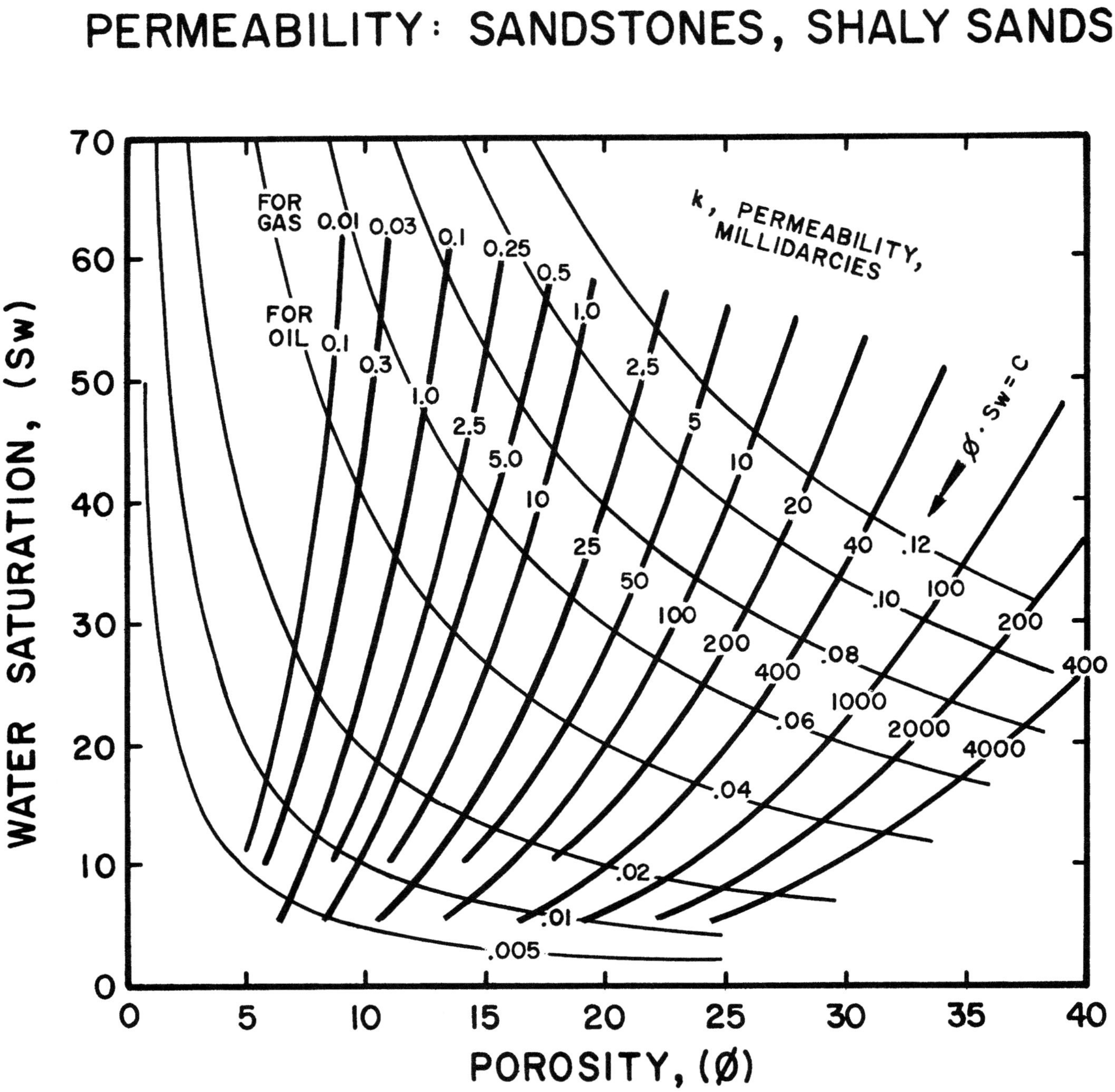

Figure 13—Cross plot of porosity (ϕ) versus water saturation (S_w), illustrating lines of equal bulk volume water (thin hyperbolic lines) and log-derived permeability (heavy vertically curved lines). (modified after: Schlumberger, 1969)

Methods of Study

Fourteen wells from the Clinton Field were selected for the study. These wells were selected because of the availability of logs, and because each one had at least three years of production history. Using a computer, calculations were made in every well at two foot intervals for porosity (ϕ), water saturation (S_w), and bulk volume water (BVW).

The computer analyzes each well by depth, porosity, water saturation and bulk volume water values, and directs the user to input porosity and bulk volume water cut-off values for each well. It next prints net feet of porosity and bulk volume water, and the net feet of porosity plus bulk volume water. These data are then used as mapping parameters.

Development and Application of Mapping Parameters

By analyzing cross plots of: 1) net porosity ($\phi \geq 10\%$) thicknesses versus cumulative upper Red Fork production (Fig. 15A); 2) net bulk volume water ($BVW \leq 0.025$) thicknesses versus cumulative upper Red Fork produc-

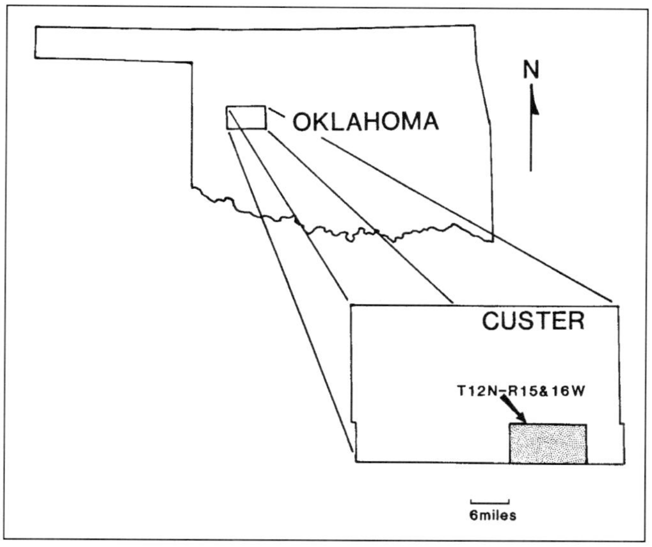

Figure 14—Index Map of Western Oklahoma showing location of the Clinton Field in Custer County, Oklahoma.

tion (Fig. 15B); and 3) net porosity (ϕ ≥12%) plus net BVW (≤0.025) thicknesses versus cumulative upper Red Fork production (Fig. 15C), we can compare the cross plots and find which one of the three net thicknesses is most sensitive to reservoir performance.

Cross plots in Figure 15 of both porosity and net BVW versus cumulative production exhibit considerable scatter (Fig. 15A and 15B). Correlation coefficients of net porosity versus cumulative production and net BVW versus cumulative production are +0.35 and +0.57, respectively (Fig. 15A and 15B). Since +1.0 is a perfect correlation, neither of these parameters appear to have a strong relationship to reservoir performance, even though net bulk volume water versus cumulative production has a better relationship to reservoir performance than does porosity. The reason for the poor correlations of net porosity and net bulk volume water to cumulative upper Red Fork production is that permeability is a function of both porosity and bulk volume water (Fig. 13).

The relationship of BVW, porosity and permeability necessitated establishing a new parameter that would be more related to reservoir performance. To meet this need, the author developed a new parameter that simultaneously relates BVW and porosity. The parameter is the net thickness of reservoir that passes certain porosity and bulk volume water cut-off values. The cut-off values for BVW and porosity are determined by examining logs from the area and checking cumulative production amounts versus different cut-offs. When values of each are established for the higher cumulative production wells, a cut-off is determined, and the cut-off then used to determine net thickness. In the Red Fork study, the cut-off values of BVW ≤ 0.025 and ϕ ≥12% were chosen by comparing several combinations of cut-offs with upper Red Fork production.**

A cross plot of net ϕ ≥12% plus BVW ≤ 0.025 versus cumulative upper Red Fork production (Fig. 15C) shows a much better agreement than did either of the other two cross plots (Fig. 15A and 15B). Also, the correlation coefficient for net ϕ ≥12% plus BVW ≤ 0.025 versus cumulative production is +0.89, much closer to +1.0 than either of the other cross plots (Fig. 15). We conclude, therefore, that the thickness of ϕ ≥12% and BVW ≤ 0.025 is a more sensitive indicator of better upper Red Fork reservoirs than is either net ϕ ≥10% or net BVW ≤ 0.025 thicknesses used alone.

Figures 16 and 17 are isopach maps; one map is of net ϕ ≥10%, the other is a map of net BVW ≤ 0.025 of the upper Red Fork sands at Clinton Field. Neither of the maps adequately outlines the better wells in the field (better wells are noted on the maps by the square and circle symbols). Figure 18, however, combines both parameters (i.e. ϕ and BVW) into one map. Notice how the better wells (square and circle symbols) are outlined by the mapping technique illustrated in Figure 18. The combined net ϕ ≥12% and BVW ≤ 0.025 map is able to define the better wells, because log-derived permeability is a

Table 6. Upper Red Fork Production Data and Net Porosity Feet.

An-Son, #1-18 Murphy, Section 18-T12N-R15W, Custer County, Oklahoma

IPF	CUM. PROD.*	ULT. PROD.	NET POROSITY
7.5 MMCF+81BO	5.88 BCF	38.0 BCF	(ϕ≥10%)
11/64" ck FTP 7850#	131,803 BO		30'

Conoco, #l-A Snider, Section 22-T12N-R16W, Custer County, Oklahoma

IPF	CUM. PROD.*	ULT. PROD.	NET POROSITY
7.6 MMCF+	2.08 BCF	11.5 BCF	(ϕ≥10%)
182 BO	56,789 BO 30'		30'
12/64" ck FTP 6750#			

An-Son, #1-20 Stidham, Section 20-T12N-R15W, Custer County, Oklahoma

IPF	CUM. PROD.*	ULT. PROD.	NET POROSITY
2.7 MMCF+	0.287 BCF	3.6 BCF	(ϕ≥10%)
38 BO	11,907 BO		30'
9/16" ck FTP 3850#			

*Cumulative production from the first three years of production.
**Remember that cut-off values for BVW and porosity will vary for different formations and different areas.

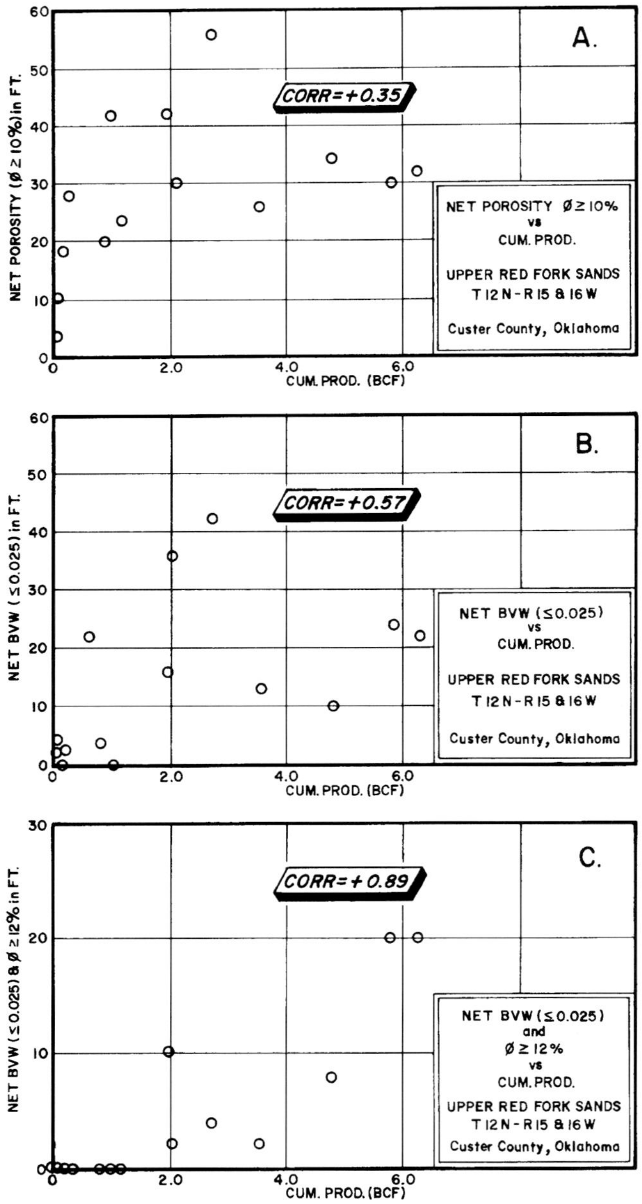

Figure 15—A. Cross plot of net porosity (ϕ ≥10%) versus cumulative production (BCF), upper Red Fork sands, Clinton Field, T12N-R15 and 16W, Custer County, Oklahoma. A correlation coefficient of only +0.35 indicates little correlation between net thickness of sand with ϕ ≥10% and cumulative production. B. Cross plot of net bulk volume water (BVW≤0.025) versus cumulative production (BCF), upper Red Fork sands, Clinton Field, T12N-R15 and 16W, Custer County, Oklahoma. A correlation coefficient of +0.57 indicates only a slightly better correlation between net thickness of sand with BVW≤0.025 and cumulative production. C. Cross plot of net bulk volume water (BVW≤0.025) and porosity (ϕ≥12%) versus cumulative production, upper Red Fork sands, Clinton Field, T12N-R15 and 16W, Custer County, Oklahoma. A correlation coefficient of +0.89 indicates a good correlation between net thickness of sand with ϕ≥12% and BVW ≤ 0.025 and cumulative production.

function of both porosity (ϕ) and bulk volume water (BVW).

Summary

Table 7 is a tabulation of production data, along with net BVW (≤0.025) plus net ϕ ≥12% and BVW ≤ 0.025. The wells listed in Table 7 are the same wells examined in Table 6. The net BVW values are very similar and yet recoverable reserves are very different. However, when recoverable reserves are compared with net ϕ ≥12% plus BVW ≤ 0.025, there is excellent agreement.

We conclude that the best parameter for mapping and delineating the good reservoirs in this formation is net thicknesses of ϕ ≥12% and BVW ≤ 0.025.

Net porosity/BVW mapping is applicable across rock types and formations, but it should only be applied to reservoirs at irreducible water saturation. The reason for this is that BVW values are invalid for estimating pore size and permeability when a reservoir is not at irreducible water saturation.

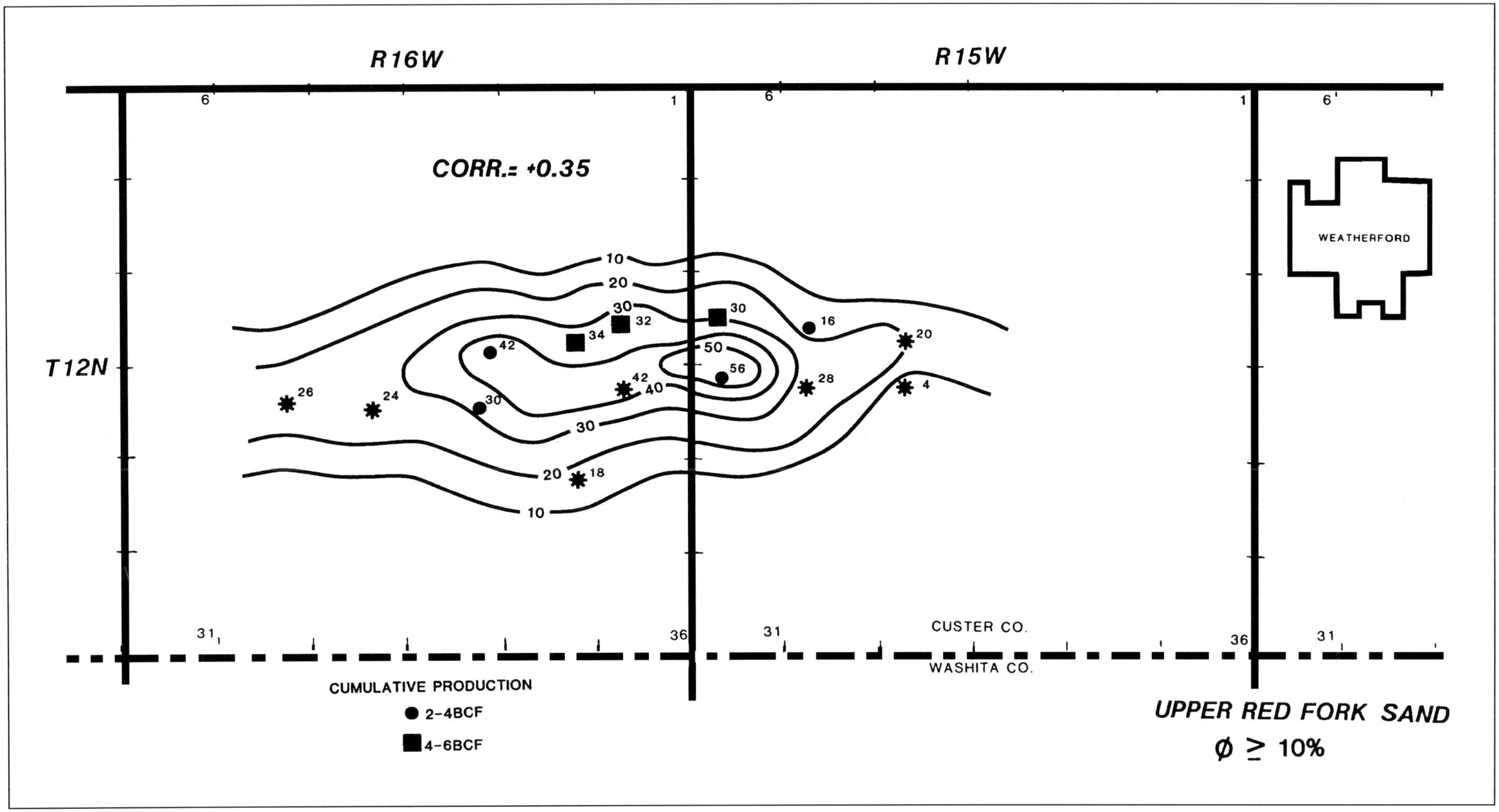

Figure 16—Isopach map of net porosity ($\phi \geq 10\%$) for the upper Red Fork sands, Clinton Field. Note that the better wells are not defined or isolated by the map, indicating that net porosity ($\phi \geq 10\%$) is a poor mapping parameter.

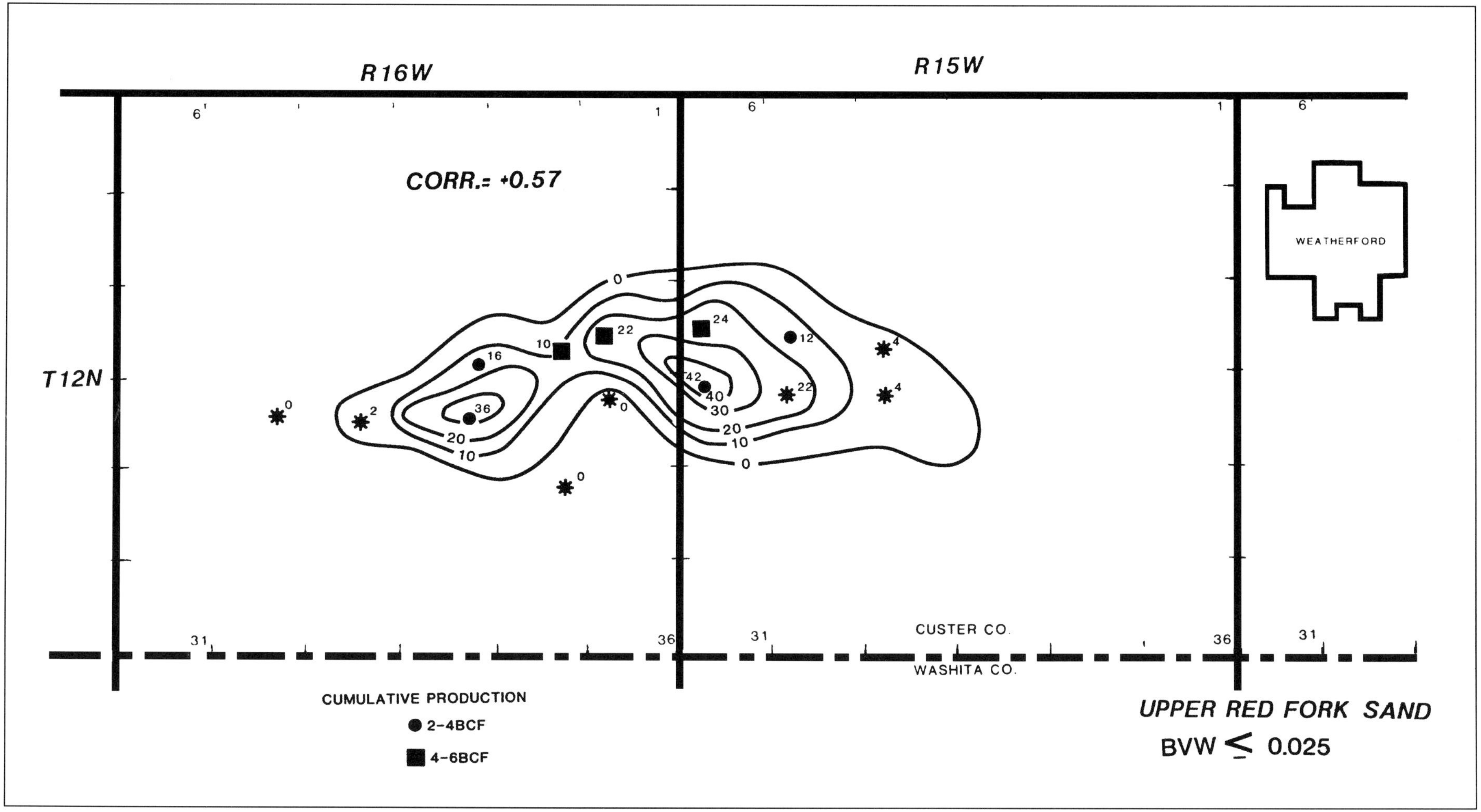

Figure 17—Isopach map of net bulk volume water (BVW ≤ 0.025) for the upper Red Fork sands, Clinton Field. Note that the map does not define or isolate the better wells, indicating that net BVW (≤ 0.025) is a poor mapping parameter.

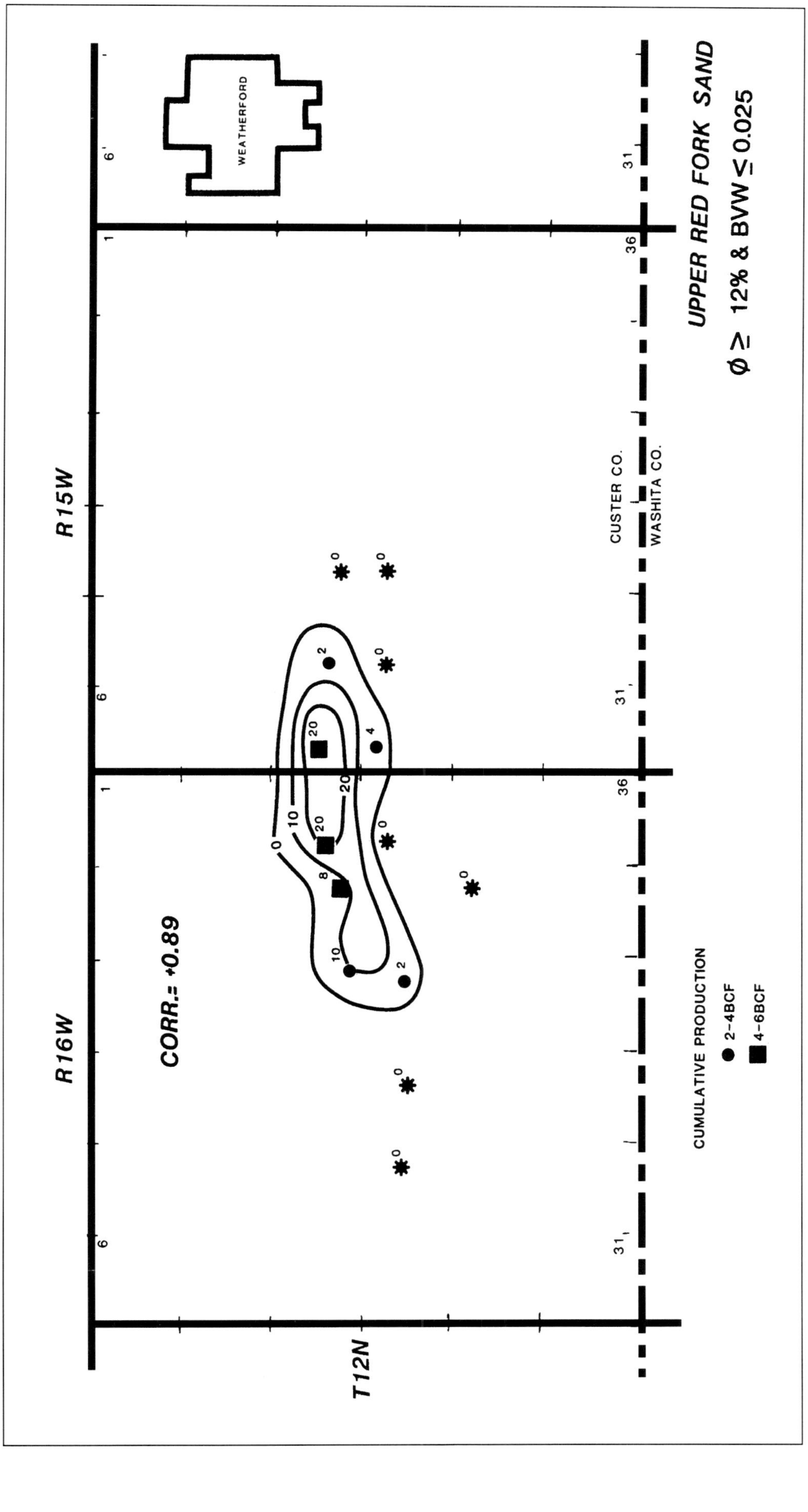

Figure 18—Isopach map of net porosity and net BVW ($\phi \geq 12\%$ and BVW ≤ 0.025) for the upper Red Fork sands, Clinton Field. Note how the better wells are defined or isolated by the map, indicating that net $\phi \geq 12\%$ and BVW ≤ 0.025 is a good mapping parameter.

Table 7. Upper Red Fork Production Data and Net BVW Feet and Net $\phi \geq 12\%$ and BVW ≤ 0.25 Feet.

An-Son, #1-18 Murphy, Section 18-T12N-R15W, Custer County, Oklahoma

IPF	CUM. PROD.*	ULT. PROD.	NET BVW (≤0.025)	NET ϕ≥12% & BVW≤0.025
7.5 MMCF+ 81 BO 11/64″ ck FTP 7850#	5.88 BCF+ 131,803 BO	38.0	24′	20′

Conoco, #l-A Snider, Section 22-T12N-R16W, Custer County, Oklahoma

IPF	CUM. PROD.*	ULT. PROD.	NET BVW (≤0.025)	NET ϕ≥112% & BVW≤0.025
7.6 MMCF 182 BO 12/64″ ck FTP 6750#	2.08 BCF+ 56,789 BO	11.5 BCF	36′	2′

An-Son, #1-20 Stidham, Section 20-T12N-R15W, Custer County, Oklahoma

IPF	CUM. PROD.*	ULT. PROD.	NET BVW (≤0.025)	NET ϕ≥112% & BVW≤0.025
2.7 MMCF+ 38BO 9/64″ ck FTP 3850#	0.287+ 11,907 BO	3.6 BCF	22′	0′

* cumulative production first three years of production

CHAPTER 5

SHALY SAND ANALYSIS CASE STUDIES

General

Five different case studies help us resolve the question posed earlier in the text, namely, if shale or clay is present in a reservoir, how do we correct log data to find the critical parameters of effective water saturation and effective porosity?

The case studies come from the San Juan basin, New Mexico; Midland basin, Texas; Anadarko basin, Oklahoma, western Kansas; and the Gulf of Mexico, and represent Mississippian, Pennsylvanian, Permian, Cretaceous and Tertiary ages.

In all of the cases, we base our assessment strategies on an unknown shale or clay distribution. Every example begins with a conventional, uncorrected log analysis; this provides some startling comparisons of different reservoir parameters, and vividly illustrates how easy it is to make costly mistakes if the time isn't taken to do shaly sand analysis in a reservoir.

Case Study 1—Pennsylvanian Cottage Grove Sandstone, Anadarko Basin

You are in the process of drilling a well in the Pennsylvanian Cottage Grove Sandstone in Woodward County, Oklahoma (Figs. 19–21). The Cottage Grove is identified as the interval from 6542 feet to 6620 feet and samples appear as fine to very fine slightly glauconitic and calcareous shaly sandstone with weak, pale yellow fluorescence. Shale beds directly above and adjacent to the Cottage Grove drilled at 5–7 mpf (minutes/foot), but the penetration rate through the sandstone was 1–3 mpf, and a 150 unit increase in gas occurred.

When you analyze the zone at 6542 feet to 6620 feet, you have available to you a log suite which includes a dual induction log (Fig. 19) and a combination neutron-density log (Fig. 20). You have gathered the following information.

$R_w = 0.05$ @ T_f

$\phi_{xp} = [(\phi_n^2 + \phi_d^2)/2]^{1/2}$

$F = 0.81/\phi^2$

$R_{mf} = 0.47$ @ T_f

$$\phi_d = \frac{\rho_{ma} - \rho_b}{\rho_{ma} - \rho_f}$$

$\rho_f = 1.0$ gm/cc

$\rho_{ma} = 2.68$ gm/cc

Conventional Analysis:

Depth (in feet)	FL	ILM	ILD	FL/ILD	ILM/ILD	R_t/R_{ILD}	R_t
6550	12	8	7	1.7	1.1	1.0	7
6566	14	8.5	6.5	2.2	1.3	0.92	6
6580	18	10	8	2.3	1.3	0.93	7
6614	13	7	7	1.9	1.0	1.0	7

Depth (in feet)	ϕ_n	ϕ_d	ϕ_{xp}	ρ_b	ϕ_d	S_w	BVW
6550	.14	.14	.14	2.47	.125	.61	0.076
6566	.145	.135	.14	2.47	.125	.66	0.083
6580	.12	.15	.135	2.45	.137	.56	0.077
6614	.145	.14	.14	2.47	.125	.61	0.076

The high water saturations (S_w) and the high bulk volume water (BVW) values determined by conventional analysis, plus the "wet" appearing resistivity profile (determined by curve separation), suggests that the Cottage Grove Sandstone is probably water productive.

However, because of the sample fluorescence (i.e. hydrocarbon show) before you give up the well you decide to do shaly sand analysis. The formulas you select to find volume of clay, effective porosity and effective water saturation are based on the log suite you have available to you.

Formulas:

$I_{GR} = (GR_{log} - GR_{min})/(GR_{max} - GR_{min})$

$V_{cl} = 0.33\,[2^{(2 \times I_{GR})} - 1.0]$ consolidated

$\phi_{nc} = \phi_n - (V_{cl} \times \phi_{nsh})$

$\phi_{dc} = \phi_d - (V_{cl} \times \phi_{dsh})$

$\phi_e = [(\phi_{dc}^2 + \phi_{nc}^2)/2]^{1/2}$

$$\phi_d = \frac{\rho_{ma} - \rho_b}{\rho_{ma} - \rho_f}$$

$$S_{we} = \frac{1}{\phi_d}\left[\sqrt{\frac{R_w}{R_t} + \left(\frac{0.35V_{cl}}{2}\right)^2} - \frac{0.35V_{cl}}{2}\right]$$

(Fertl, 1975)

$q = (\phi_{xp} - \phi_e)/\phi_{xp}$

Given:

$\rho_{ma} = 2.68$ g/cc

$\rho_f = 1.0$ g/cc

Determined:

$\phi_{nsh} = .28$ @ 6626′ (Fig. 20)

$\phi_{dsh} = .08$ @ 6626′ (Fig. 20)

GR_{max} = 105 @ 6626′ (Fig. 20)
GR_{min} = 15 @ 6786′ (Fig. 22)

The Fertl (1975) formula is selected as the shaly sand method, because the adjacent shales are as resistive or have more resistivity (see Chapter 3) than the Cottage Grove Sandstone .

Shaly Sand Analysis:

Depth (in feet)	ρ_b (Fig. 21)	ϕ_d*	GR_{log}	I_{GR}	V_{cl}
6550	2.47	.125	53	.53	.26
6566	2.47	.125	45	.33	.19
6580	2.45	.137	45	.33	.19
6614	2.47	.125	60	.50	.33

* This is the density porosity (ϕ_d) you use in the Fertl (1975) formula.

Depth (in feet)	ϕ_{nc}	ϕ_{dc}	ϕ_e	S_{we}	BVW	q
6550	.067	.119	.097	.404	.039	.31
6566	.092	.119	.106	.511	.054	.24
6580	.067	.135	.107	.42	.045	.21
6614	.053	.114	.089	.357	.032	.36

As part of the analysis, you cross plot effective porosity (ϕ_e) and q on the Shaly Sand Producibility chart (Fig. 23) to find out whether or not the well falls in a producible area on the chart.

The results of the shaly sand analysis on the Cottage Grove Sandstone indicate that the water saturations are lower (S_{we} = 36% to 51%) than the non-clay corrected water saturations (S_{wt} = 56% to 66%). Along with the lower water saturations, there is a concomitant decrease in the clay corrected bulk volume water values (BVW = 0.032 to 0.054), compared with the non-clay corrected bulk volume water values (BVW = 0.076 to 0.083). The lower clay corrected S_{we} and BVW data suggest that the Cottage Grove Sandstone should be productive. The effective porosity (ϕ_e) and q data for two of the depths (6566′ and 6580′), plot along the line separating productive from non-productive (Fig. 23) regions. The other two depths plot in the non-productive region (Fig. 23). We conclude from our shaly sand analysis that the Cottage Grove Sandstone should be productive, but that the production will be marginal.

The Cottage Grove Sandstone was perforated from 6544 feet to 6618 feet and had an initial production of 1.68 MMCFGPD plus 1 BOPD. After two years the well had cumulative production of 0.231 BCF and 601 BO.

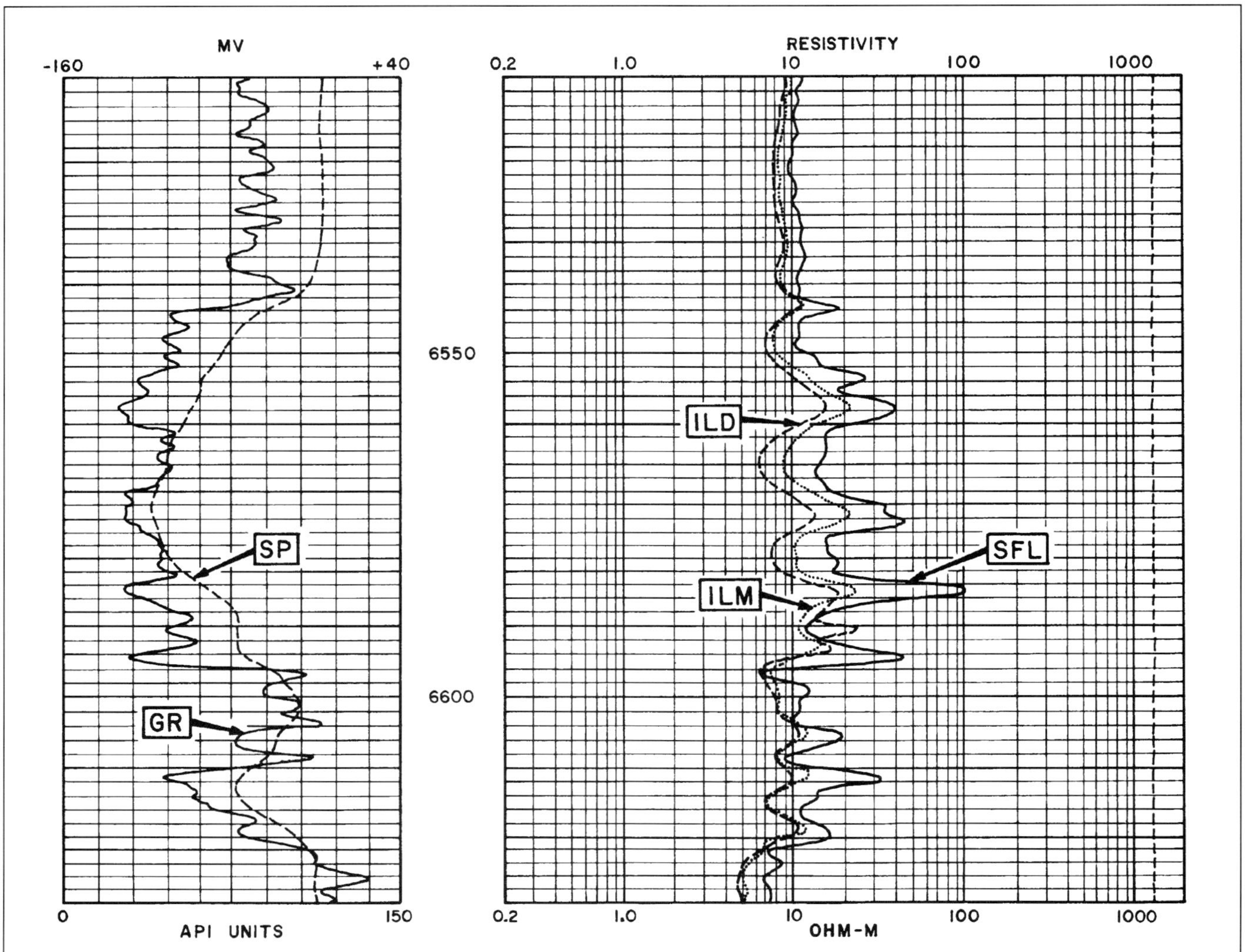

Figure 19—Dual induction—SFL log through the Pennsylvanian Cottage Grove Sandstone, western Oklahoma. The Cottage Grove Sandstone is the interval from 6542 feet to 6620 feet.

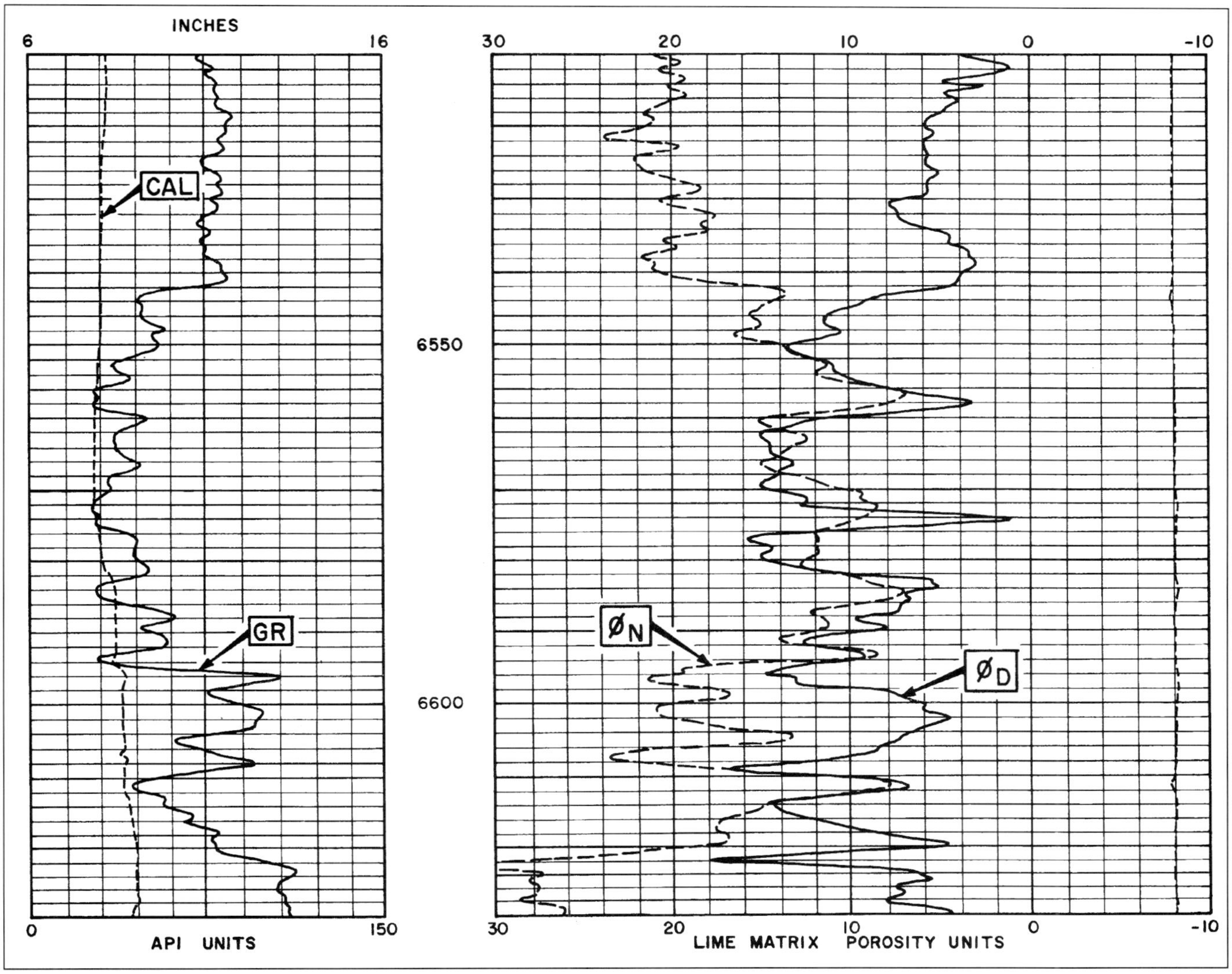

Figure 20—Gamma ray neutron-density log through the Pennsylvanian Cottage Grove Sandstone, western Oklahoma. The Cottage Grove Sandstone is the interval from 6542 feet to 6620 feet.

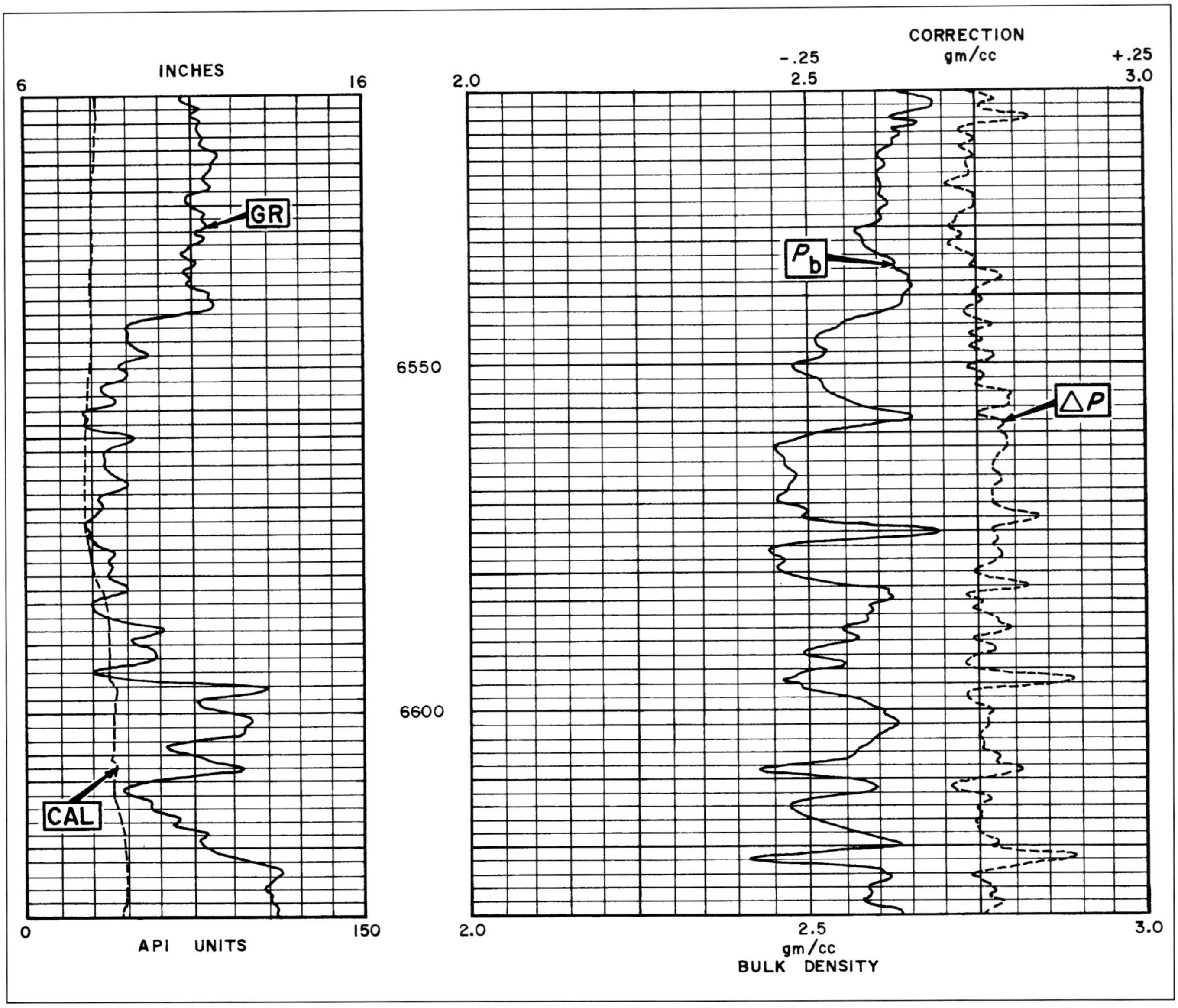

Figure 21—Gamma ray-density log through the Pennsylvanian Cottage Grove Sandstone western Oklahoma. The Cottage Grove Sandstone is the interval from 6542 feet to 6620 feet.

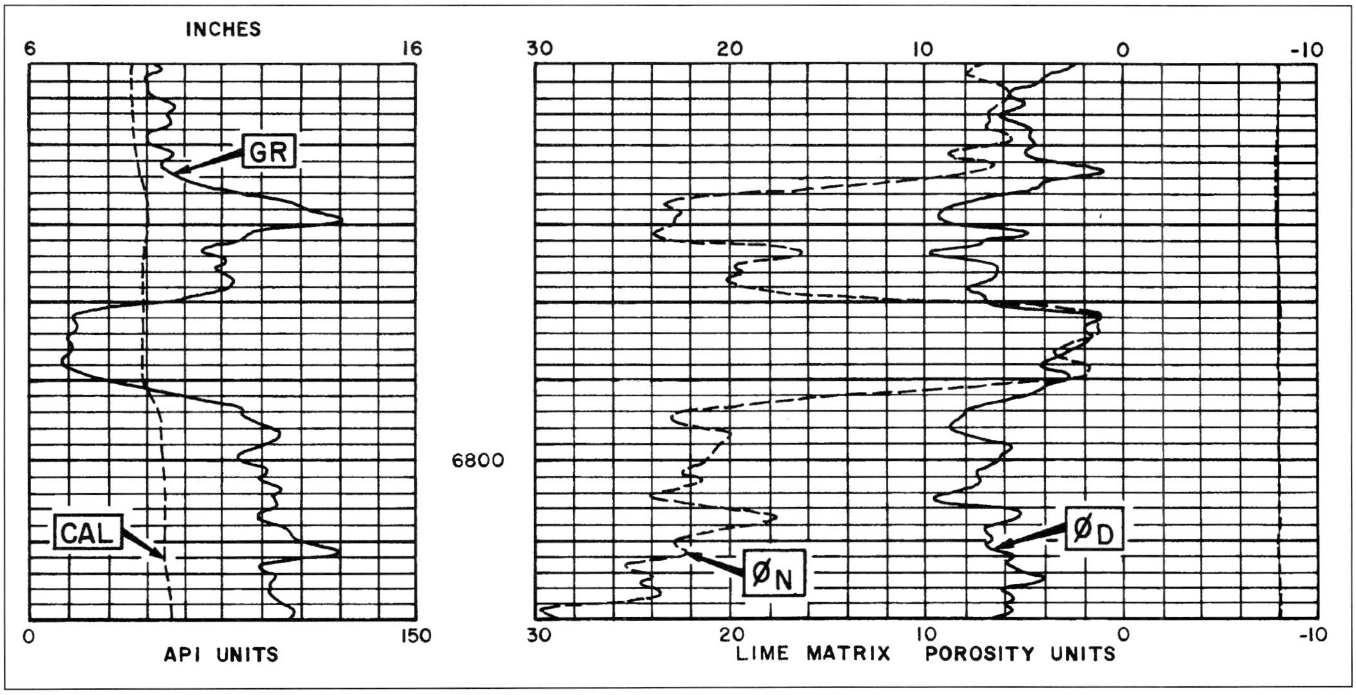

Figure 22—Gamma ray neutron-density log through a clean non-porous sand (6780′ – 6792′) in the same well as figures 19 to 21.

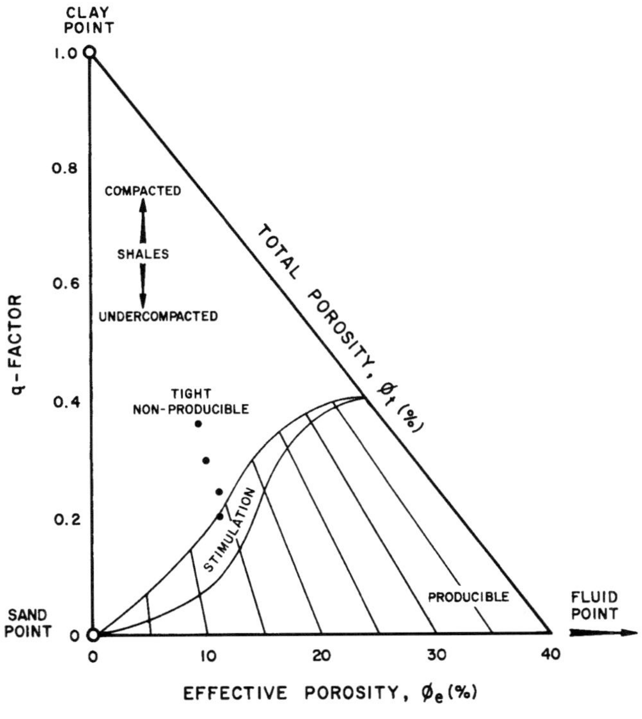

Figure 23—Shaly sand producibility chart of q versus effective porosity (ϕ_e) for the Cottage Grove Sandstone in western Oklahoma.

Case Study 2—Cretaceous Pictured Cliffs Sandstone, San Juan Basin

Your company has just reached total depth (TD) in a Cretaceous Pictured Cliffs exploratory well several miles basinward (NE) of the most basinward Pictured Cliffs production in the San Juan Basin. The samples taken from the Pictured Cliffs Sandstone at a depth of 3678 feet to 3802 feet (Fig. 24) appeared as light-grey, fine to very fine grained sandstone, slightly calcareous, with visible amounts of gray shale.

No fluorescence was noted on the samples but there was a 45 unit gas increase as the Pictured Cliffs Sandstone was drilled.

Conventional log analysis gives the following information:

Given:

$R_w = 0.2 @ T_f$

$F = 0.81/\phi^2$

$\phi_{xp} = [(\phi_n^2 + \phi_d^2)/2]^{1/2}$

$$\phi_d = \frac{\rho_{ma} - \rho_b}{\rho_{ma} - \rho_f}$$

Conventional Analysis:

Depth (in feet)	ϕ_n	ϕ_d*	ϕ_{xp}	R_t	S_w	BVW
3684	.13	.17	.151	19	.54	0.092
3710	.15	.11	.132	19	.84	0.092
3716	.15	.10	.128	17	.98	0.098
3728	.13	.13	.130	20	.69	0.090

* This density porosity can also be used in the Fertl (1975) formula because the CNL-FDC log was run on a sandstone matrix.

The high water saturations (S_w) and high bulk volume water (BVW) values suggest the reservoir is water productive, but understanding, as you do, the ability of clay to affect water saturation, you realize the need for a shaly sand analysis. Your first step is to construct a neutron-density porosity cross plot, which you use to determine both the volume of clay in the Pictured Cliffs Sandstone and the range of effective porosities (Fig. 25). Because the neutron and density porosities are recorded on a sand matrix (2.65 g/cc), the clean sand line (Fig. 25) is drawn from 0.0 neutron porosity and 0.0 density porosity so that both neutron and density porosities, plotted on the clean sand line, are equal. The shale point represents the neutron and density porosity readings from the shale zone below the Pictured Cliffs sand at a depth of 3842 feet (Fig. 24; $\phi_{nsh} = 21$ and $\phi_{dsh} = 2$). You then construct a line from the shale point to 100 percent neutron-density porosity. The clean sand line is scaled in porosity using the neutron and density porosity values on the x and y axes. The percentage of clay is scaled by subdividing the line between the clean sand point at 0.0 porosity and the shale point into equal subdivisions. Clay percentages are then extrapolated to 100 percent porosity and porosity percentages are drawn parallel to the percent clay line (Fig. 25). You plot the neutron and density porosities from the Pictured Cliffs Sandstone (Fig. 24; 3678′ to 3802′) and determine the following:

Range of clay percent 0.0 to 22% (Fig 25)

Range of effective porosities 10% to 16% (Fig. 25)

The formulas you use for shaly sand analysis are, once again, determined by the log suite, which includes neutron-density and resistivity logs.

Formulas:

$I_{GR} = (GR_{log} - GR_{min})/(GR_{max} - GR_{min})$

$V_{cl} = 0.33\ [2^{(2 \times I_{GR})} - 1.0]$ consolidated

$\phi_{nc} = \phi_n - (V_{cl} \times \phi_{nsh})$

$\phi_{dc} = \phi_d - (V_{cl} \times \phi_{dsh})$

$\phi_e = [(\phi_{nc}^2 + \phi_{dc}^2)/2]^{1/2}$

$$S_{we} = \frac{1}{\phi_d}\left[\sqrt{\frac{R_w}{R_t} + \left(\frac{0.35V_{cl}}{2}\right)^2} - \frac{0.35V_{cl}}{2}\right]$$

(Fertl, 1975)

S_{we} = Dual Water*

$q = (\phi_{xp} - \phi_e)/\phi_{xp}$

The Fertl (1975) method was selected, because the Pictured Cliffs Sandstone resistivity (R_t) is almost equal to the adjacent shale resistivity (R_{sh}). You are concerned that the Fertl (1975) method may be unreliable, because the Pictured Cliffs sandstone's formation water resistivity ($R_w = 0.2$) is greater than $0.1\Omega - M$ (see: Chapter 3). You decide that the high R_w value necessitates the use of the Dual Water method along with the Fertl (1975) method, as a means for checking the accuracy, in this instance, of the Fertl method.

Determined:

$GR_{min} = 40 @ 3640'$ (Fig. 24)

$GR_{max} = 140 @ 3842'$ (Fig. 24)

$\phi_{nsh} = .21 @ 3842'$ (Fig. 24)

$\phi_{dsh} = .02 @ 3842'$ (Fig. 24)

$R_{sh} = 16\Omega - M @ 3842'$ (Fig. 24)

$R_{cl} = 0.4 \times R_{sh}$ (Fertl and Hammack, 1971)

$R_{cl} = 0.4 \times 16$

$R_{cl} = 6.4\ \Omega - M$ dispersed clay resistivity

Shaly Sand Analysis:

Depth (in feet)	GR_{log}	I_{GR}	V_{cl}	ϕ_{nc}	ϕ_{dc}	ϕ_e	q
3684	64	.24	.13	.103	.167	.139	.08
3710	80	.40	.25	.098	.105	.102	.23
3716	85	.45	.29	.089	.094	.092	.28
3728	70	.30	.17	.094	.127	.112	.14

Volumes of clay values are calculated from the gamma ray log instead of the neutron-density log, because the Pictured Cliffs Sandstone is a gas-bearing formation (see: Chapter 3).

Shaly Sand Analysis:

Depth (in feet)	S_{we} (Fertl)	S_{we} (Dual H_2O; R_{sh})	S_{we} (Dual H_2O; R_{cl})	BVW
3684	.484	.623	.513	.067
3710	.616	.713	.476	.063
3716	.690	.799	.505	.063
3728	.574	.708	.541	.064

* For a review of the Dual Water equations see Chapter 3.

You note that in the above calculations, the S_{we} (Dual H_2O; R_{sh}) is much greater than S_{we} (Fertl), suggesting the Pictured Cliffs Sandstone has dispersed clay (see: Chapter 3). S_{we} (Fertl) and S_{we} (Dual H_2O; R_{cl}) values are closer together, and are less than S_{we} (Dual H_2O; R_{sh}). Therefore, either the Fertl (1975) or Swe (Dual H_2O; R_{cl}) methods can be used for S_{we}. The Fertl (1975) method is valid because R_{sh} and R_{cl} (R_{sh} = 16; R_{cl} = 6.4) are much greater than R_w (R_w = 0.2), and the S_{we} (Dual H_2O; R_{cl}) method is valid because the clay is dispersed (see Chapter 3). After you have calculated both effective porosity (ϕ_e) and q then cross plot the data on the Shaly Sand Producibility chart (Fig. 26).

The application of shaly sand analysis to the Pictured Cliff Sandstone results in both a lower water saturation (S_{we} = 48 to 69%) and lower bulk volume water values (0.063 to 0.067), compared with the non-clay corrected values (S_w = 0.54 to 98%) and BVW = 0.092 to 0.10). You determine that the Pictured Cliffs Sandstone should, therefore, be productive. The neutron-density cross plot indicates that at depths 3684 feet and 3728 feet the clay content is very low* (Fig. 25). The other depths (3710′ and 3716′), however, have much higher clay content (Fig. 25). The Shaly Sand Producibility chart for the Pictured Cliffs Sandstone (Fig. 26) reveals that the sands at 3684 feet and 3728 feet both plot in the productive region, but the other depths (3710′ and 3716′) plot in the non-productive region. Thus, both the neutron-density cross plot (Fig. 25) and the Shaly Sand Producibility chart (Fig. 26) indicate that the best reservoirs in the Pictured Cliffs Sandstone are the zones at 3684 feet and 3728 feet.

The Pictured Cliffs Sandstone was perforated from 3680 feet to 3733 feet, and had an initial production of 2.872 MMCFGPD with no water on a 3/4-inch choke with a shut-in casing pressure (SICP) of 1110 psi. The cumulative production after six months was 0.057 BCF with a production of 322 MCFPDG.

* The volumes of clay determined by the neutron-density cross plot (Fig. 25) are probably too low, because the presence of gas shifts the data points toward the clean sand line (i.e., lower V_{cl}).

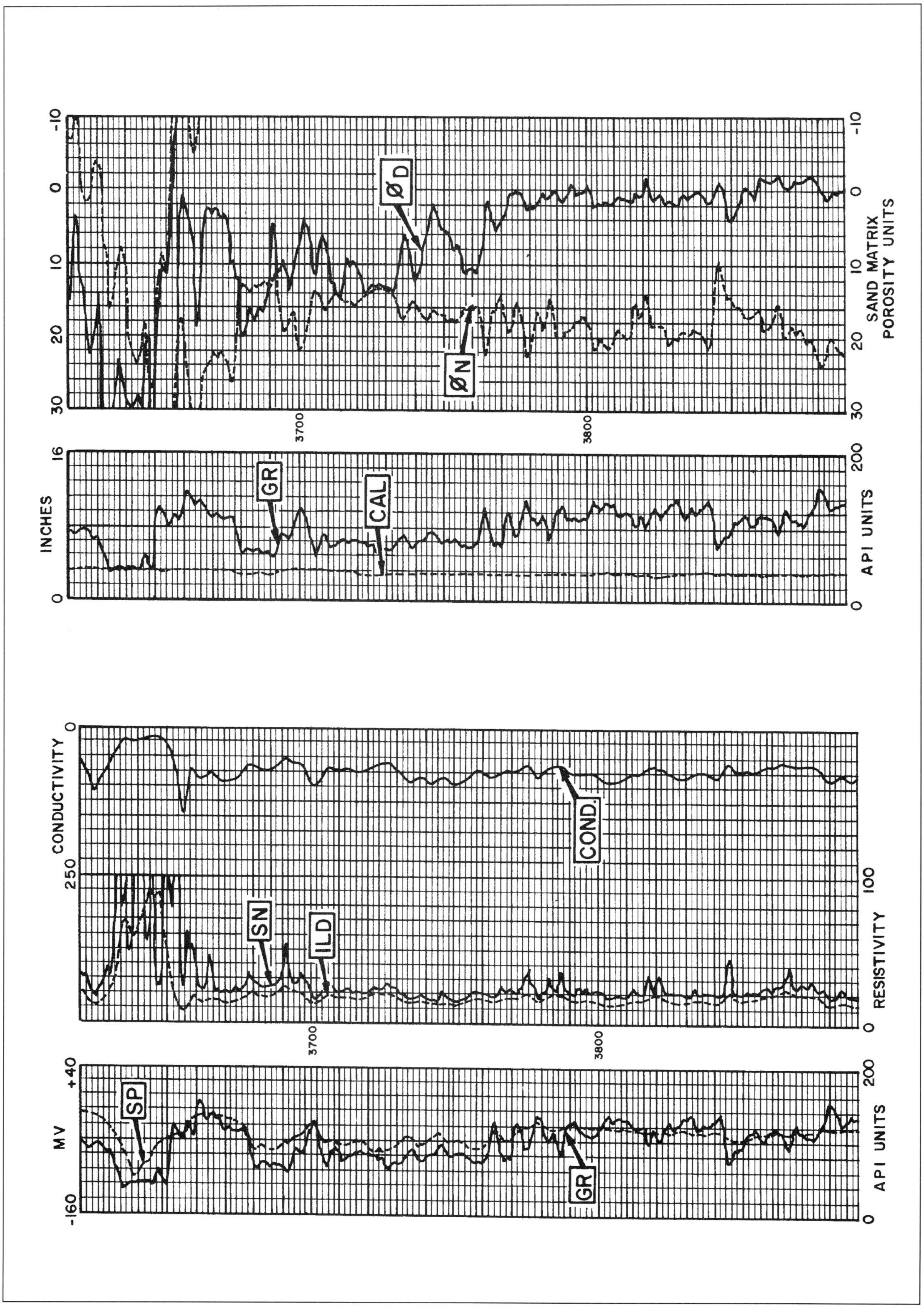

Figure 24—Induction-short normal log with SP plus a gamma ray neutron-density log through the Cretaceous Pictured Cliffs Sandstone, San Juan Basin, New Mexico. The Pictures Cliffs Sandstone is the interval from 3678 feet to 3802 feet.

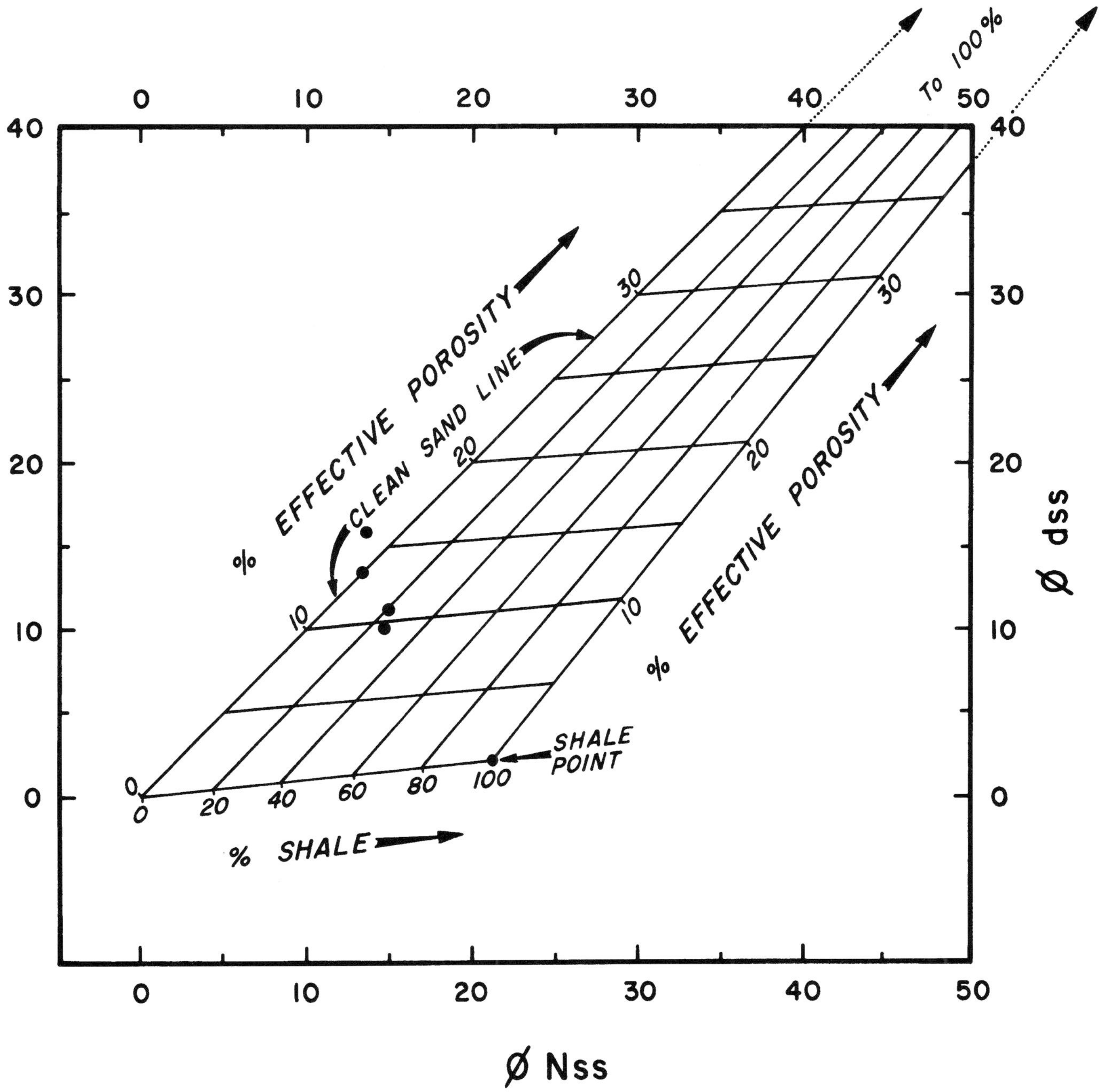

Figure 25—Neutron and density cross plot chart for determining volume of clay and effective porosity for the Pictured Cliffs Sandstone, San Juan Basin, New Mexico.

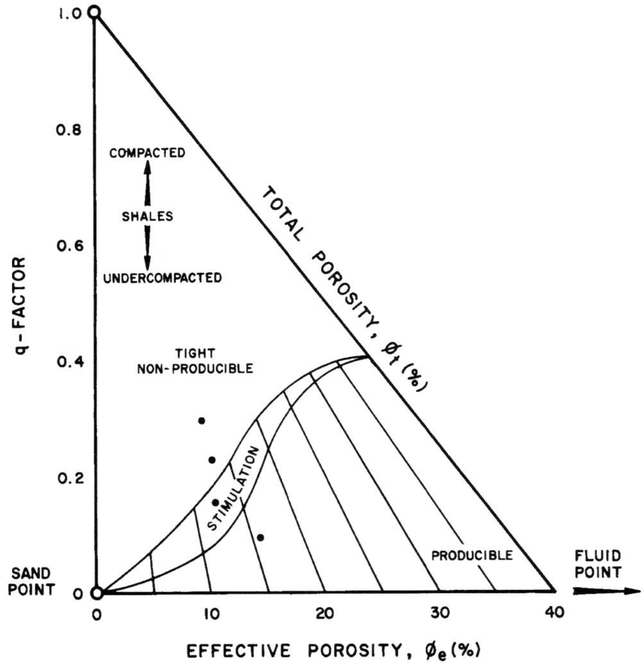

Figure 26—Shaly Sand Producibility chart of q versus effective porosity (ϕ_e) for the Pictured Cliffs Sandstone San Juan Basin, New Mexico.

Case Study 3—Permian Spraberry Sandstone, Midland Basin

While drilling a Spraberry development well in the Midland Basin, you encounter the deep water Permian Spraberry Sandstone (Figs. 27–29) at a depth of 7720 feet to 7750 feet. A sample description of the Spraberry sandstone from 7720 feet to 7750 feet reads as follows:

Gray, poorly sorted, fine-grained calcareous shaly sandstone with good light-blue fluorescence and weak streaming cut.

A conventional log analysis of this zone produces the following:

Given:

$R_w = 0.04 @ T_f$
$F = 0.81/\phi^2$
$\rho_{ma} = 2.68$ g/cc
$\rho_f = 1.0$g/cc
$\phi_{xp} = (\phi_n + \phi_d) / 2$

Conventional Analysis:

Depth (in feet)	ϕ_n	ϕ_d	ϕ_{xp}	ρ_b	ϕ_d	R_t	S_w	BVW
7724	.22	.18	.20	2.39	.17	2.6	.66	0.112
7732	.23	.23	.23	2.32	.21	2.9	.50	0.106
7738	.25	.24	.245	2.30	.23	2.0	.55	0.127

The high water saturations (S_w) and the very high bulk volume water (BVW) values concern you because you feel that the zone may be wet. Also, the high bulk volume water values worry you because even if the reservoir is not wet, the high BVW values indicate that the effective permeability may be low, due to the high amount of bound water in the porosity. Therefore, you realize the Shaly Sand Producibility chart will be very important to your final decision about the commercial producibility of the well.

In order to do shaly sand analysis you select these equations:

$I_{GR} = (GR_{log} - GR_{min})/(GR_{max} - GR_{min})$
$V_{cl} = 0.33\,[2^{(2 \times I_{GR})} - 1.0]$ consolidated
$\phi_{nc} = \phi_n - (V_{cl} \times \phi_{nsh})$
$\phi_{dc} = \phi_d - (V_{cl} \times \phi_{dsh})$
$\phi_e = (\phi_{nc} + \phi_{dc})/2$

$$S_{we} = \frac{1}{\phi_d}\left[\sqrt{\frac{R_w}{R_t} + \left(\frac{0.35 \times V_{cl}}{2}\right)^2} - \frac{0.35 \times V_{cl}}{2}\right]$$

$$\phi_d = \frac{\rho_{ma} - \rho_b}{\rho_{ma} - \rho_f}$$

$$q = (\phi_{xp} - \phi_e) / \phi_{xp}$$

The Fertl (1975) method is chosen because the resistivities of the adjacent shales are greater than the resistivity of the Spraberry Sandstone.

Given:

$\rho_{ma} = 2.68$ g/cc
$\rho_f = 1.0$ g/cc
$GR_{min} = 20$

Determined:

$\phi_{nsh} = .27$ @ 7760′ (Fig. 28)
$\phi_{dsh} = .17$ @ 7760′ (Fig. 28)
$GR_{max} = 135$ @ 7760′ (Fig. 28)

Shaly Sand Analysis:

Depth (in feet)	ρ_b (Fig. 29)	ϕ_d*	GR_{log}	I_{GR}	V_{cl}
7724	2.39	.17	75	.48	.31
7732	2.32	.21	68	.42	.26
7738	2.30	.23	65	.39	.24

*This is the density porosity you use in the Fertl (1975) formula.

Depth (in feet)	ϕ_{nc}	ϕ_{dc}	ϕ_e	S_{we}	BVW	q
7724	.136	.127	.132	.477	.063	.34
7732	.160	.186	.173	.383	.066	.25
7738	.185	.199	.192	.459	.088	.22

Because of your concern for effective permeability, you cross plot effective porosity (ϕ_e) and q on the Shaly Sand Producibility chart (Fig. 30).

After shaly sand analysis, the water saturations (S_{we} = 38% to 48%) and bulk volume water values (BVW = 0.063 to 0.088) are lower than the non-clay corrected values (S_w = 50% to 66% and BVW = 0.11 to 0.13), thus indicating the Spraberry Sandstone should be productive. Your major concern is reservoir permeability, because the original (non-clay corrected) BVW values were very high (BVW = 0.11 to 0.13). These high BVW values indicate a high bound water content, which means effective porosity (ϕ_e) and permeability could be very low. The Shaly Sand Producibility chart (Fig. 30) for the Spraberry Sandstone reveals that two of the depths (7732′ and 7738′) fall in the producible region (Fig. 30). Therefore, depths 7732 feet and 7738 feet should have sufficient porosity and permeability to be productive. The Spraberry Sandstone was perforated from 7722 feet to 7746 feet and had an initial pumping production of 236 BOPD plus 63 MCFGPD and 11 barrels of load water. After five years, the lease, which consists of three wells, had a cumulative production of 160,000 BO and 0.042 BCF.

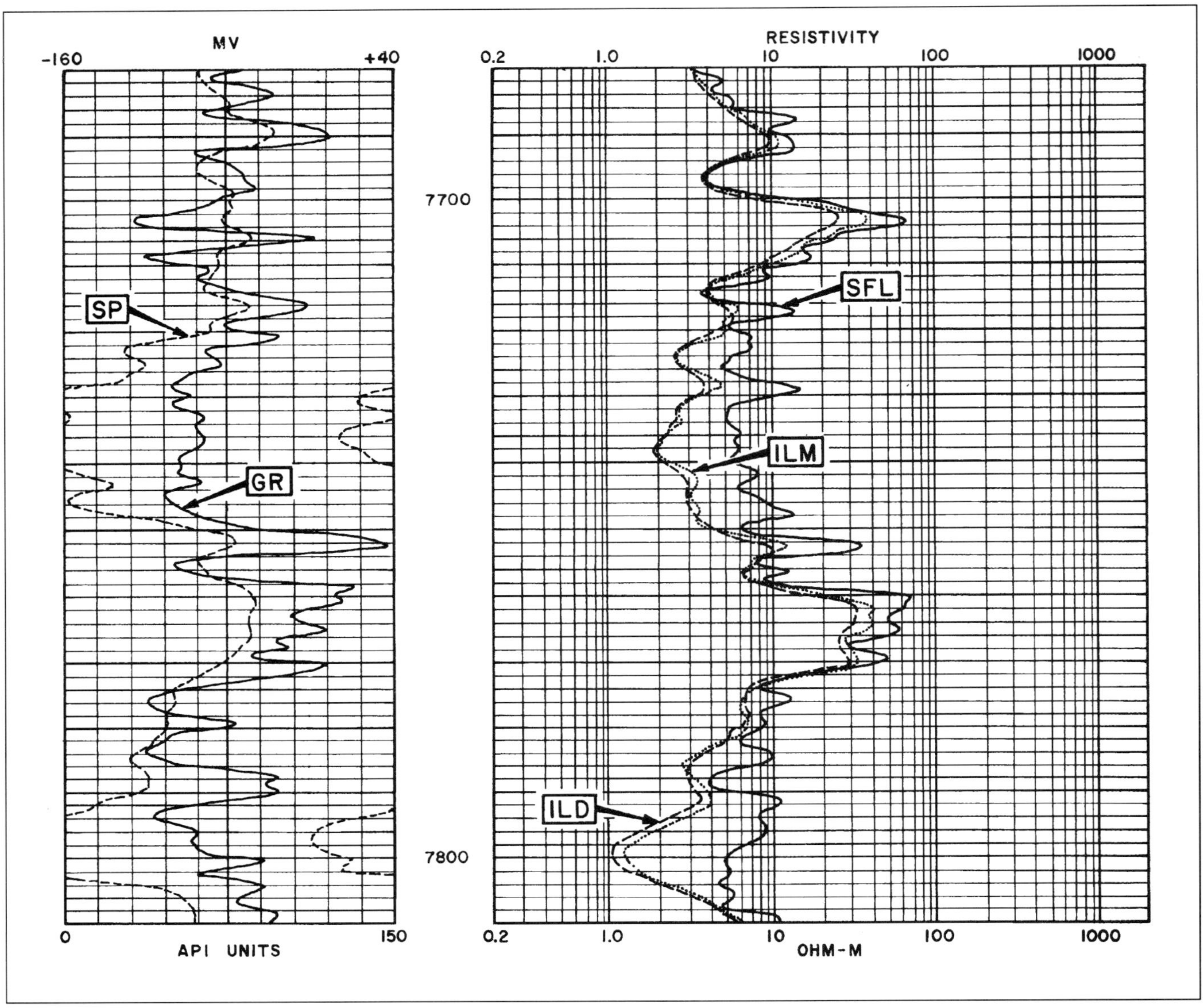

Figure 27—Dual Induction—SFL log through the Permian Spraberry Sandstone, Midland Basin. The Spraberry sand interval of interest is from 7720 feet to 7750 feet.

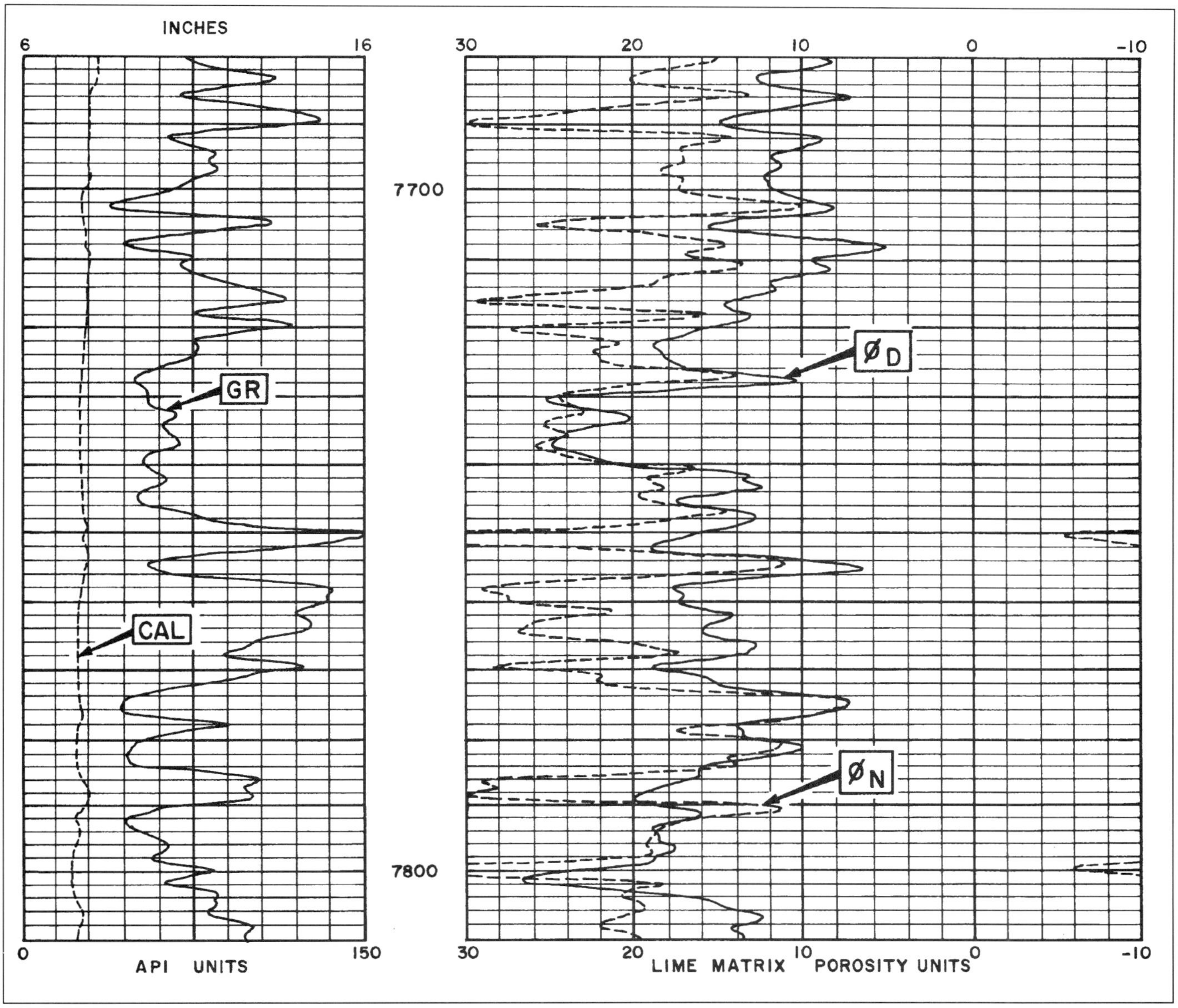

Figure 28—Gamma ray neutron-density log through the Permian Spraberry Sandstone, Midland Basin. The Spraberry sand of interest is from 7720 feet to 7750 feet.

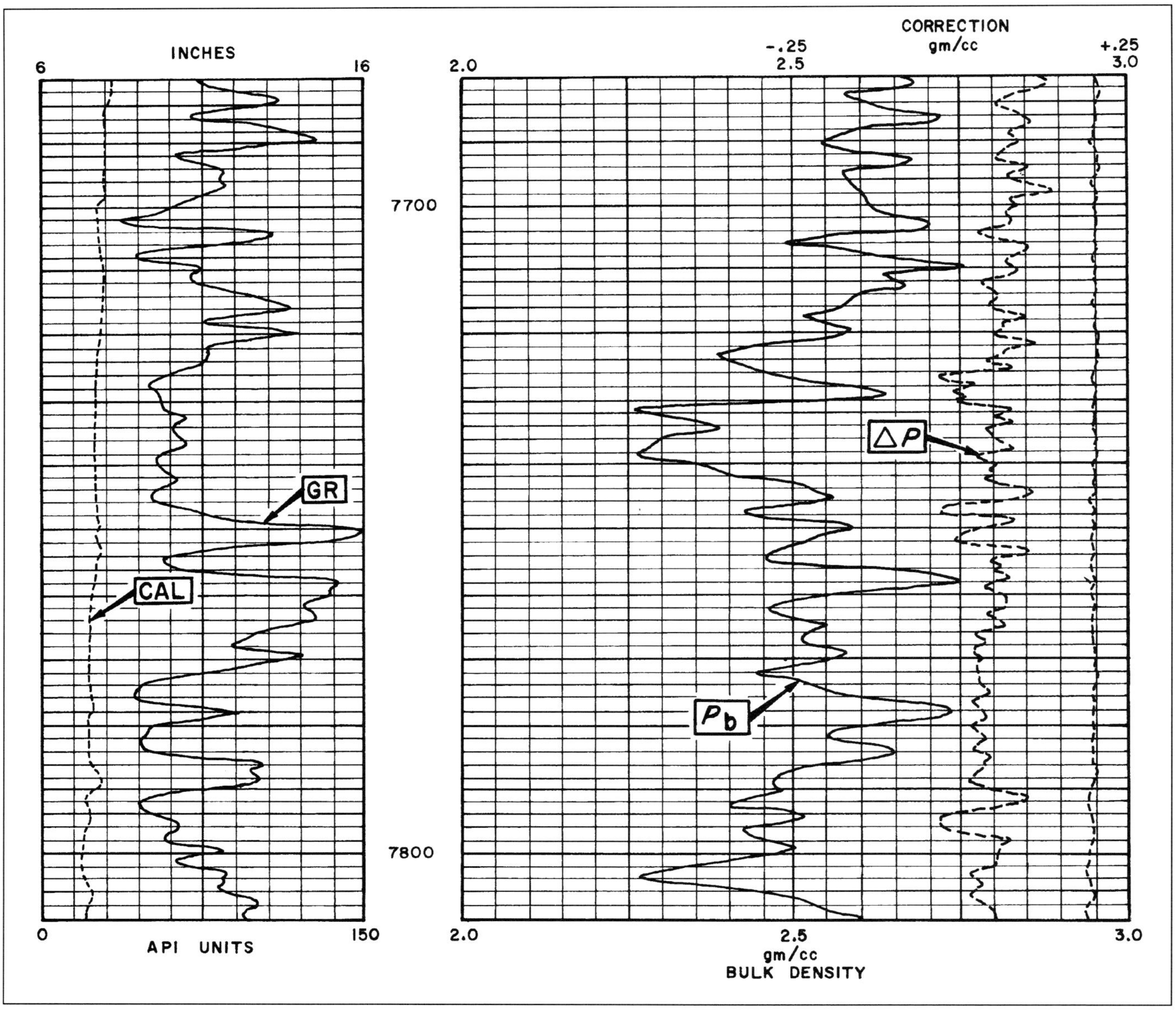

Figure 29—Gamma ray density log through the Permian Spraberry Sandstone, Midland Basin. The Spraberry sand of interest is from 7720 feet to 7750 feet.

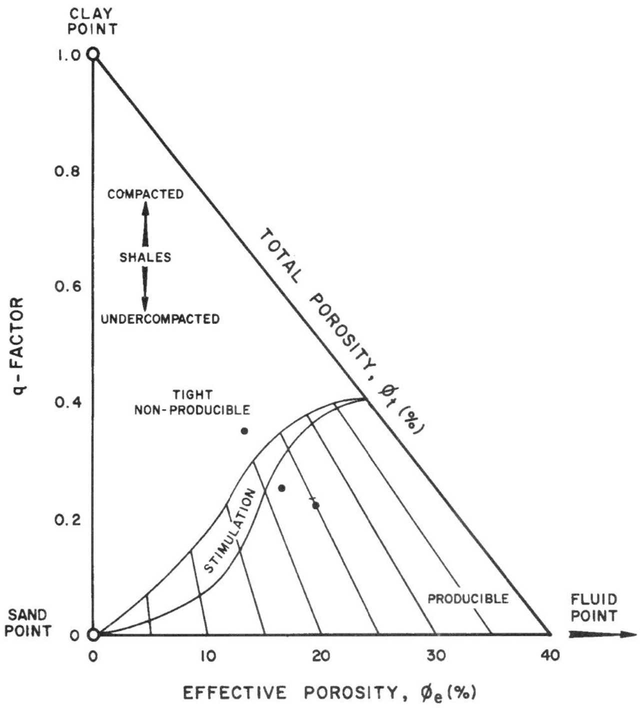

Figure 30—Shaly sand producibility chart of q versus effective porosity (ϕ_e) for the Spraberry Sandstone, Midland Basin, Texas.

Case Study 4—Plio-Pleistocene Sand, Offshore Gulf of Mexico

Your company drills a well, which penetrates through an interval from 9724 feet to 9836 feet in a Plio-Pleistocene sand located offshore Gulf of Mexico. You can see on the resistivity log (Fig. 31) that this sand has very low resistivity (< 1.0Ω – M), suggesting that the interval may be water-productive. Conventional log analysis, using density porosity (ϕ_d), yields water saturations of 55 percent to 60 percent. The sample description of sidewall cores taken in the interval from 9726 feet to 9836 feet is very fine grained shaly sand, slightly calcareous. SEM analysis reveals that clay is present as grain coatings, sometimes entirely filling the pore space (Vajnar, et al., 1977). The logging suite for the well is an electric ISF-sonic log (Fig. 31) plus a gamma ray density log (Fig. 32). Therefore, the shaly sand equation you select to use is the Dispersed Clay method (Alger et. al., 1963). You have assembled the following information:

$R_w = 0.02$ @ T_f

$GR_{min} = 45$

$\phi_{ma} = 2.68$ gm/cc

$\Delta t_{ma} = 56$ µsec/ft

$\rho_f = 1.0$ gm/cc

$\Delta t_f = 189$ µsec/ft.

The formulas necessary to do shale sand analysis by the Dispersed Clay method are:

$I_{GR} = (GR_{log} - GR_{min})/(GR_{max} - GR_{min})$

$V_{cl} = 0.083\,[2^{(3.7 \times I_{GR})} - 1.0]$ unconsolidated

$$\phi_e = \left(\frac{\rho_{ma} - \rho_b}{\rho_{ma} - \rho_f}\right) \times V_{cl}\left(\frac{\rho_{ma} - \rho_{sh}}{\rho_{ma} - \rho_f}\right)$$

$$\phi_s = \frac{\Delta t_{log} - \Delta t_{ma}}{\Delta t_f - \Delta t_{ma}} \times \frac{100}{\Delta t_{sh}}$$

$$\phi_d = \frac{\rho_{ma} - \rho_b}{\rho_{ma} - \rho_f}$$

$$q = (\phi_s - \phi_d)/\phi_s$$

$$S_{we} = \left[\sqrt{\frac{0.8}{\phi_s^2} \times \frac{R_w}{R_t} + \left(\frac{q}{2}\right)^2} - \left(\frac{q}{2}\right)\right] / (1 - q)$$

(Alger et al., 1963)

Determined:

$GR_{max} = 90$ @ 9700′ (Fig. 32)

$\Delta t_{sh} = 118$ @ 9700′ (Fig. 31)

$\rho_{sh} = 2.43$ @ 9700′ (Fig. 32)

Shaly Sand Analysis:

Depth (in feet)	GR_{log}	I_{GR}	V_{cl}
9730	55	.22	.06
9750	60	.33	.11
9770	65	.44	.17
9790	58	.29	.09

Depth (in feet)	Δt	ϕ_s	ρ_b	ϕ_d	q	ϕ_e
9730	108	.33	2.28	.24	.27	.23
9750	111	.35	2.26	.25	.29	.23
9770	115	.38	2.25	.26	.32	.23
9790	119	.40	2.23	.27	.33	.26

Depth (in feet)	R_t	S_{we}	BVW
9730	.9	.399	.092
9750	.8	.40	.092
9770	.7	.395	.091
9790	.75	.352	.092

You remember to plot your effective porosity (ϕ_e) and q values on the Shaly Sand Producibility chart (Fig. 33) to determine if the shale has drastically reduced permeability. Remember that the non-clay-corrected water saturations were 55 to 60 percent. The lower water saturations (S_{we} = 35% to 40%) and the low bulk volume water values (BVW = 0.091 to 0.092) obtained after shaly sand analysis tell you that the low resistivity (<1Ω – M) Plio-Pleistocene sand should be productive. Effective porosity (ϕ_e) and q data, when plotted on the Shaly Sand Producibility chart (Fig. 33), cluster well within the producible region. Therefore, the volume of shale (V_{cl} = 6% to 17%) in the Plio-Pleistocene sand is insufficient to drastically reduce porosity and permeability.

The Plio-Pleistocene sand illustrated in Figures 31 and 32 was perforated from 9725 feet to 9805 feet and 9813 feet to 9833 feet. The well initially flowed 1,305 BOPD and 1.05 MMCFGPD on a 15/64-inch choke with no water. After two years of production the well was still producing 598 BOPD and had a cumulative production of 291,551 BO and no water.

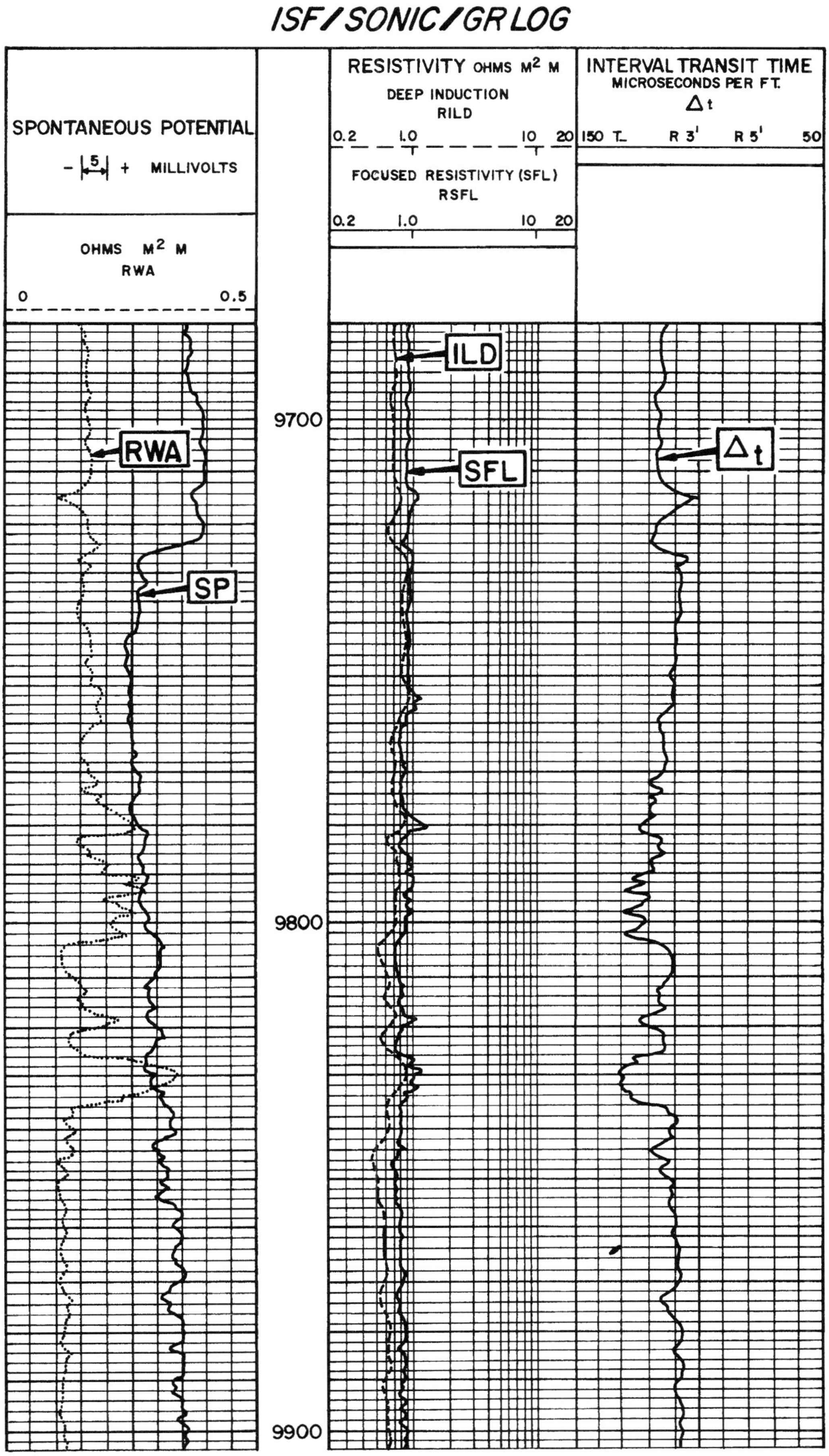

Figure 31—ISF—sonic log through a Plio-Pleistocene sand, Gulf Coast. The Plio-Pleistocene sand interval is from 9724 to 9836 feet. Note that the sand has very low (<1.0W – M) resistivity.

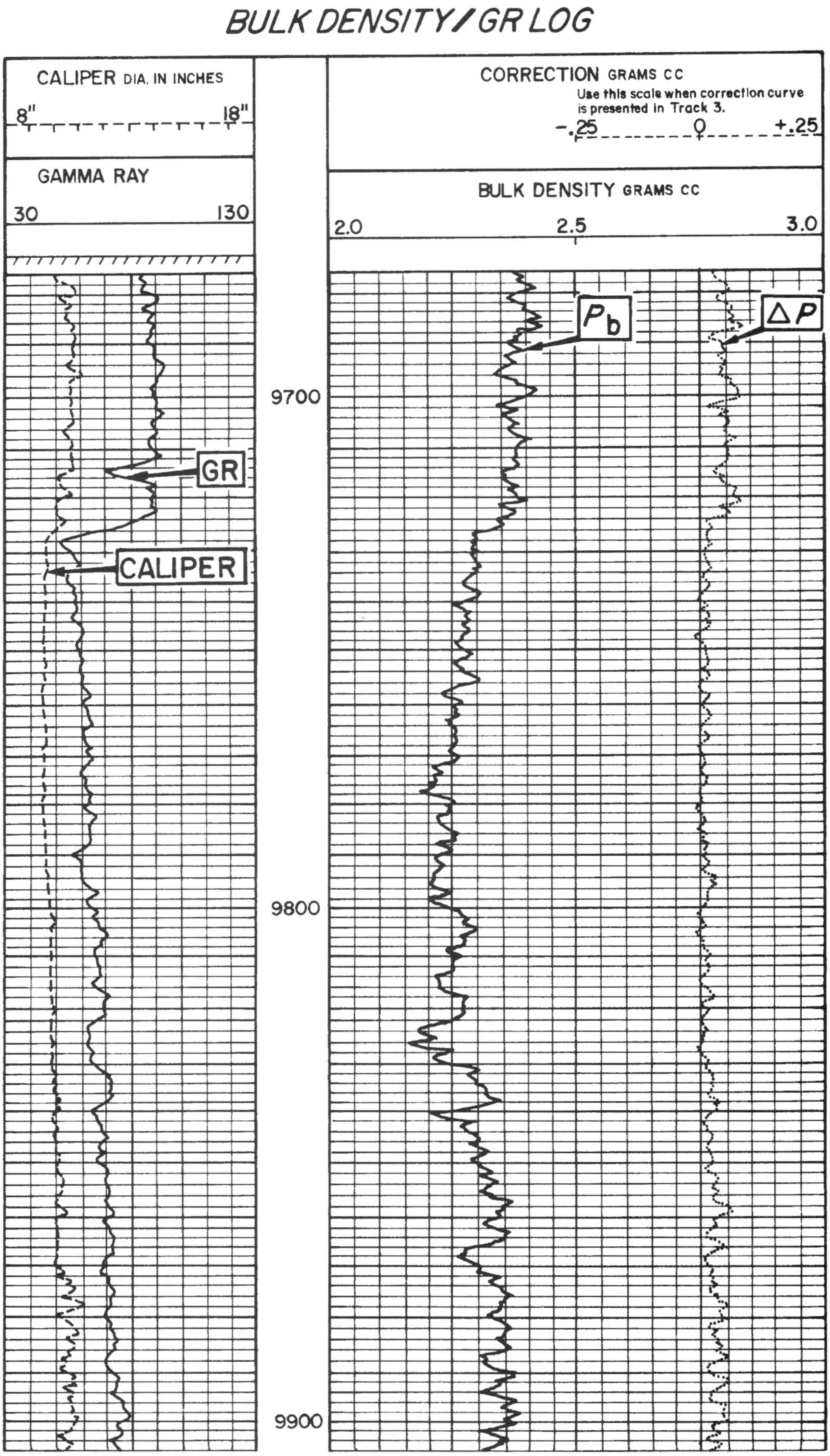

Figure 32—Gamma ray density log through a Plio-Pleistocene sand, Gulf Coast. The Plio-Pleistocene sand interval is from 9724 feet to 9836 feet.

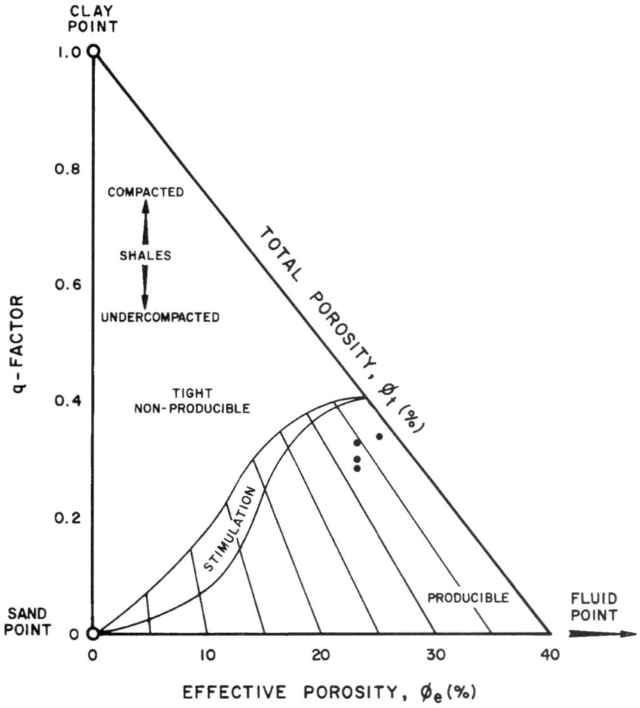

Figure 33—Shaly Sand Producibility chart of q versus effective porosity (ϕ_e) for the Plio-Pleistocene sand, Gulf Coast.

Case Study 5—Mississippian Chester Sandstone, Southwestern Kansas

This last problem is offered as an opportunity for exercising some creative thinking. It reminds us that log analysis is still a dynamic, evolving discipline.

Figures 34 and 35 are a dual induction log and combination neutron-density log through two Mississippian Chester sandstones (5704′–5718′ and 5724′–5738′) in western Kansas. Note that both of the sand intervals have low gamma ray readings (15–30 GAPI), indicating a very low shale or clay content. On the dual induction log (Fig. 34), the upper sandstone (5704′–5718′) has resistivities ranging from 19Ω – M to 38Ω – M, and the lower sandstone 2.5 Ω – M to 4Ω – M. The neutron-density log (Fig. 35) indicates that the porosities of the two sands are very similar. A conventional log analysis of these two Chester sandstones is shown below:

Given:

$R_w = 0.04$ @ T_f

$R_{mf} = 0.65$ @ T_f

$F = 0.81/\phi^2$

$\phi_{xp} = (\phi_n + \phi_d)/2$

Conventional Log Analysis:

Depth (in feet)	ϕ_n	ϕ_d	ϕ_{xp}	R_t	S_w	BVW
5706	.09	.14	.115	38	.254	0.029
5712	.12	.16	.14	20	.287	0.040
5728	.135	.15	.143	3.2	.704	0.101
5736	.135	.18	.158	2.5	.721	0.114

The very high water saturations (S_w = 70–72%) and very high bulk volume water (BVW = 0.101–0.114) in the lower sand (5724′–5738′) confirm what was suspected by the much lower resistivities in the lower sand, namely, that there is an oil/water contact between the upper and lower sands.

A logical question to ask is whether the lower sand could be shaly, causing the resistivity to be lowered. Yet, the gamma ray log indicates both sands have low shale or clay content by the low gamma ray response on both the upper and lower sands (Fig. 34). The SP log also indicates low shale or clay content in both sands by the strong SP response (Fig. 34). Therefore, we conclude the lower sand must be wet.

An examination of the EPT (electromagnetic propagation tool) log (Fig. 36) indicates both sands are hydrocarbon-bearing, because EPT (EMCP) porosity (water-filled porosity) is less than neutron-density (PHIA) porosity in both the upper and lower sands.

The water saturation of the flushed zone (S_{xo}) can be calculated using the EMCP (EPT) porosity and PHIA (neutron-density) porosity to determine if the water saturations in the flushed zone (S_{xo}) are similar in the upper and lower sands.

S_{xo} from EPT (EMCP) and neutron-density (PHIA) porosity (Fig. 36):

Depth (in feet)	ϕ_{PHIA}	ϕ_{EMCP}	S_{xo} ($S_{xo} = \phi_{EMCP}/\phi_{PHIA}$)
5706	.12	.06	.50
5712	.14	.06	.43
5728	.14	.06	.43
5736	.16	.07	.44

The EPT log (Fig. 36) indicates both zones should be productive.

SEM analysis of the lower sand indicates very thin illite clay coatings on the sand grains. The illite coatings on the grains increase surface conductance and irreducible water saturation (S_{wirr}), which results in the decrease in resistivity (Fig. 34). SEM analysis of the upper sands reveals no clay coatings.

The amount of clay, volumetrically, is very small so that the gamma ray and SP logs cannot detect it (Fig. 34). Therefore, a straight forward shaly sand approach does not work because V_{cl} is too low to make a sufficient correction on the calculated water saturation.

The type of analysis this shaly reservoir needs is the analysis technique that employs the single textural parameter "W" (Guillotte et. al., 1979), where "W" is used for both m and n in the Archie water saturation equation:

$$S_w = \frac{1}{\phi} \times \left(\frac{R_w}{R_t}\right)^{1/W}$$

(Guillotte et. al., 1979)

The "W" parameter is calculated using the EPT log and shallow resistivity log by the equation:

$W = [\log(R_z/R_{sfl})] / [\log(\phi_{EMCP})]$

where:

R_{sfl} = shallow resistivity log (Fig. 34)

ϕ_{EMCP} = EPT or water-filled porosity (Fig. 36)

$$R_z = \frac{1}{\dfrac{z}{R_w} + \dfrac{1 - z}{R_{mf}}}$$

R_z = mixed water resistivity

where:

z = 0.075 intergranular porosity

z = 0.035 vuggy porosity

If we apply the "W" method to determine the water saturation of the lower sand (Fig. 1: 5724′–5738′), we find:

Depth (in feet)	R_z	R_{sfl}	ϕ_{EMCP}	W	S_{we}
5728	.303	15	.06	1.39	.30
5736	.303	17	.07	1.51	.41

The water saturation calculated with the "W" parameter is the effective water saturation (S_{we}). The effective porosity (ϕ_e) is calculated by the following method:

$S_b = S_{wt} - S_{we}$

$\phi_e = \phi_t - (\phi_t \times S_b)$

where:

S_b = clay bound water saturation

S_{wt} = total water saturation uncorrected for clay, calculated earlier

S_{we} = effective water saturation corrected for clay by the "W" technique
ϕ_e = effective porosity
ϕ_t = total porosity

Next, we calculate the effective porosity, q and BVW of the lower sand (Fig. 34: 5724′–5738′).

Depth (in feet)	S_{wt}	S_{we}	S_b	ϕ_e	BVW	q
5728	.704	.30	.40	.09	.027	.37
5736	.721	.41	.31	.11	.045	.30

$q = (\phi_{xp} - \phi_e)/\phi_{xp}$

Now cross plot q and ϕ_e for the lower sand (5728′ and 5736′) on the Shaly Sand Producibility chart (Fig. 37).

Due to the very low volume of clay in the Chester sandstone from 5724 feet to 5738 feet, conventional shaly sand analysis was impossible because V_{cl} was too low to make sufficient correction. Using the single textural parameter "W" for both m and n in the Archie equation for the lower of the two Chester sandstones (5724′ to 5738′; Figs. 34 and 35), yields effective (clay-corrected) water saturations of 30 and 41 percent, thus verifying the presence of hydrocarbons detected by the EPT log (Fig. 36). When the neutron-density porosity (ϕ_{xp} = 14% to 16%) is corrected to effective porosity (ϕ_e = 9% to 11%) in the lower sandstone, the porosity is reduced. The loss in effective porosity is due to the clay bound water.

The Shaly Sand Producibility chart (Fig. 37) indicates that the lower Chester sandstone may be marginally productive, because only one of the two depths (5736 feet) plots near the productive region.

The two Chester sandstones were perforated from 5700 feet to 5738 feet (gross perforated interval) and the well had an initial production of 157 BOPD. After 90 days the well had a cumulative production of 9,000 BO plus 230 BW.

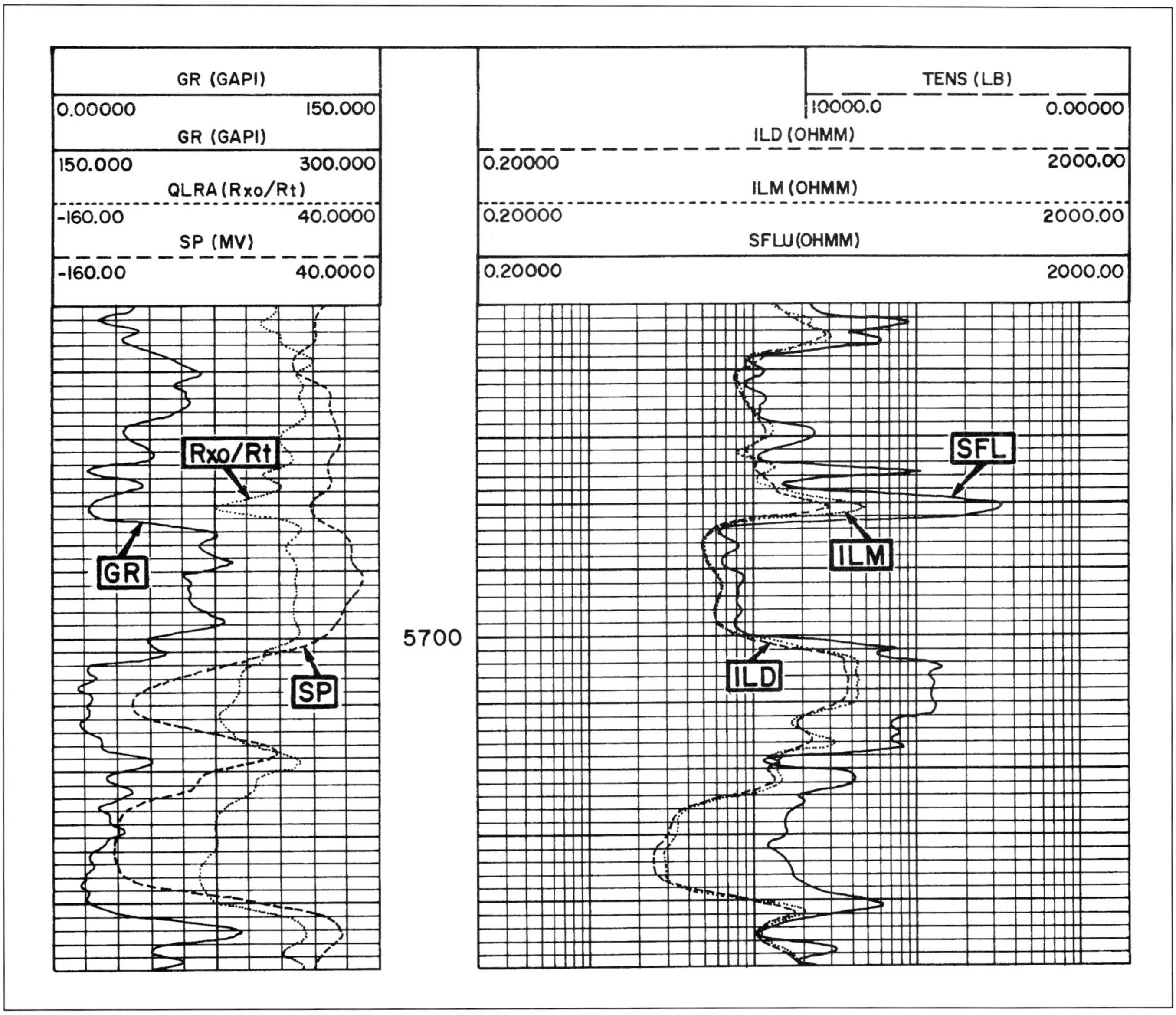

Figure 34—Dual induction—SFL log through two Mississippian Chester sandstones in western Kansas. The Chester sandstones are from 5704 feet to 5738 feet.

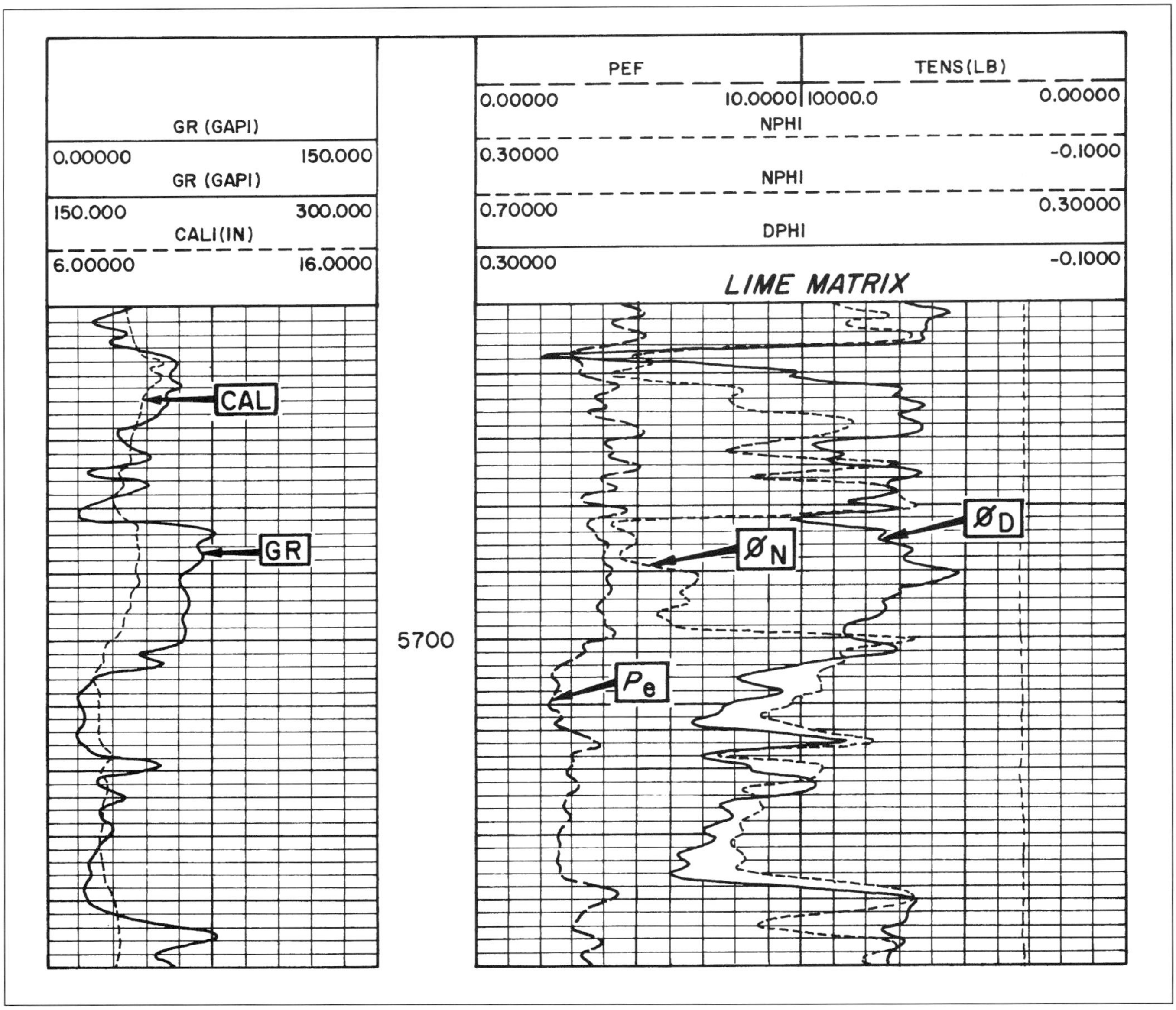

Figure 35—Gamma ray neutron-density log through two Mississippian Chester sandstones in western Kansas. The Chester sandstones are from 5704 feet to 5738 feet.

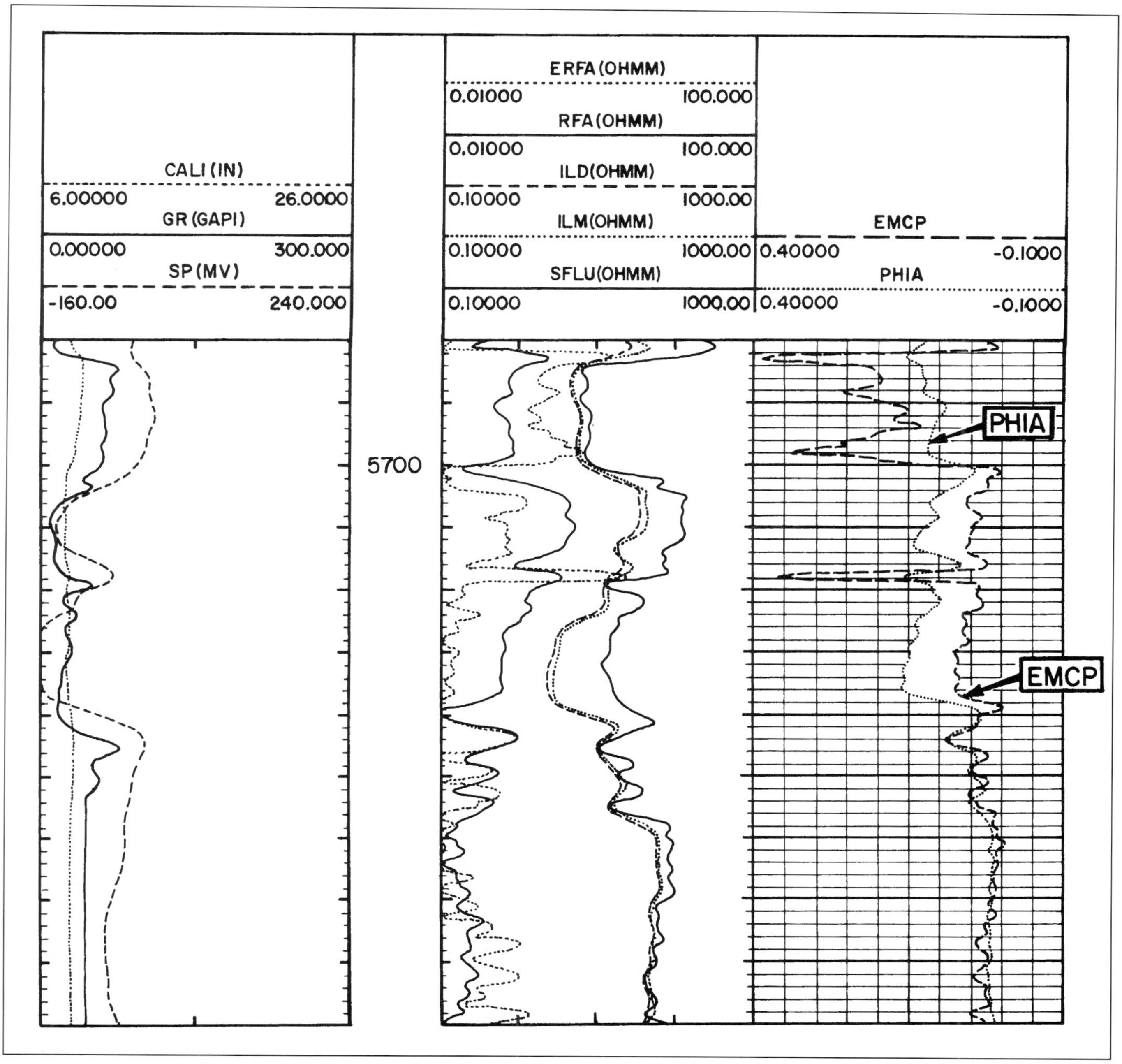

Figure 36—Computer-generated neutron-density porosity (ϕ_{PHIA}) and EPT porosity (ϕ_{EMCP}) log through two Mississippian Chester sandstones in western Kansas. The Chester sandstones are from 5704' to 5738'. Note that in both the upper sand (5704'–5718') and the lower sand (5724'–5738'), the EPT porosity (EMCP) is less than neutron-density porosity (PHIA), indicating that both the sands should be productive.

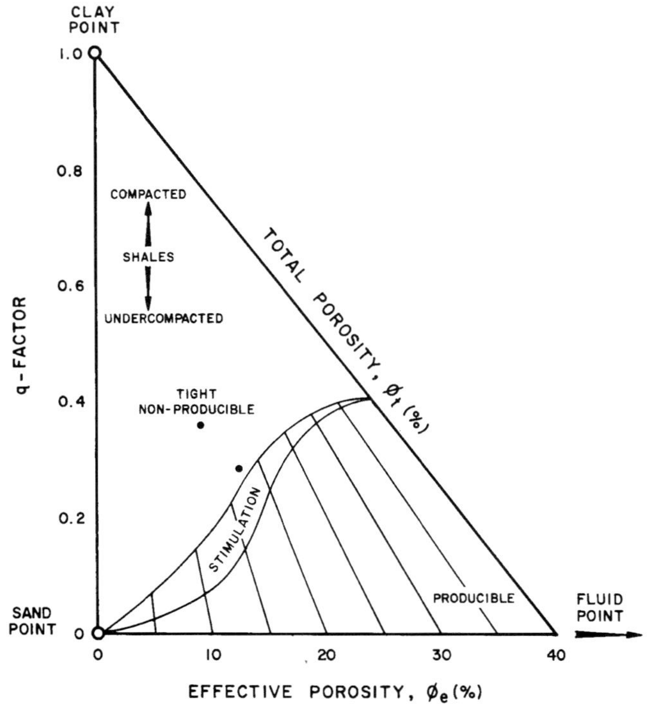

Figure 37—Shaly Sand Producibility chart of q versus effective porosity (ϕ_e) for lower Chester Sandstone (5724' – 5738') in western Kansas.

REFERENCES CITED

Ali, S. A., 1981, Sandstone Diagenesis: application to hydrocarbon exploration and production: Gulf Science and Technology Co., 221 p.

Almond, W. R., 1979, A geologic appreciation of shaly sands: Soc. of Professional Well Log Analysts, 20th Annual Logging Symposium Trans., Paper WW.

Alger, R. P., L. L. Raymer, W. R. Hoyle, and M. P. Tixier, 1963, Formation density log applications in liquid filled holes: Jour. of Petroleum Technology SPE-AIME (March), p. 310-380.

Berry, W. R. IV, 1982, Shale/Clay Properties *in* Shaly Sand: Soc. of Professional Well Log Analysts, Reprint Volume, Houston, Texas p. II-I3.

Clavier, C., G. Coates, and J. Dumanoir, 1977, The theoretical and experimental bases for the 'dual water' model for the interpretation of shaly sands: Soc. of Professional Engineers of AIME, Denver, SPE paper 6859.

Dresser Atlas, 1979, Log Interpretations Charts: Houston, Dresser Industries Inc., 107 p.

Dresser Atlas, 1983, Log Interpretation Charts: Houston, Dresser Industries Inc., 149 p.

Dewan, J. T., 1983, Essentials of Modern Open-Hole Log Interpretation: PennWell Publishing Co., Tulsa, Oklahoma, 361 p.

Elphick, R. Y., 1988, Log analysis of an induction electric log: Geobyte, vol. 3, no. 3, p. 70-72.

Fertl, W. H., 1975, Shaly sand analysis in development wells: Soc. of Professional Well Log Analysts, 16th Annual Logging Symposium Trans., paper A.

Fertl, W. H., and G. W. Hammack, 1971, A comparative look at water saturation computations in shaly pay sands: Soc. of Professional Well Log Analysts, 12th Annual Logging Symposium Trans., Paper R.

Fertl, W. H., and W. C. Vercellino, 1978, Predict water cut from well logs, *in* Practical log analysis - 4: Oil and Gas Jour., (May 15, 1978 - Sept. 19, 1979).

Fertl, W. H., 1982, Introduction *in* Shaly Sand: Soc. of Professional Well Log Analysts, Reprint Volume, Houston, Texas.

Guillotte, J. G., J. Schrank, and E. Hunt, 1979, Smackover reservoir: interpretation case study of water saturation versus production: Gulf Coast Assoc. of Geological Societies, Transactions v. 29, p. 121-126.

Hilchie, D. W., 1982, Advanced Well Log Interpretation: Golden Colo., D. W. Hilchie Inc.

Hilchie, D. W., 1978, Applied Open-Hole Log Interpretation: Golden Colo., D. W. Hilchie Inc.

Johnson, W. L., and W. A. Linke, 1977, Some practical applications to improve formation evaluation of sandstones in the Mackenzie Delta: Canadian Well Log Society, 6th Annual Logging Symposium Trans., paper R.

Lang, W. H., 1982, Shaly Sand: Soc. of Professional Well Log Analysts, Reprint volume Houston, Texas.

Munson, T. W., 1988, Depositional, diagenetic, and production history of the Upper Morrow Buckhaults Sandstone Farnsworth field, Ochiltree County, Texas: M.S. Thesis, West Texas State Univ. 201 p.

Neashan, T. W., 1977, The morphology of dispersed clay in sandstone reservoirs and its effects on sandstone shalinesss, pore space, and fluid flow properties: Soc. of Petroleum Engineers of the AIME, Denver, SPE paper 6858.

Patchett, J. G., 1975, An investigation of shale conductivity: Soc. of Professional Well Log Analysts, 16th Annual Logging Symposium Trans. Paper U.

Patchett, J. G. and D. C. Herrick, 1982, A Review of Saturation Models *in* Shaly Sand: Soc. of Professional Well Log Analysts, Reprint volume, Houston, Texas, p. IIII-III7..

Poupon, A., C. Clavier, J. Dumanoir, R. Gaymond, and A. Mish, 1970, Log analysis of sand-shale sequences—a systematic approach: Jour. of Petroleum Technology, (July), p. 867-881.

Poupon, A., J. Levlaux, 1971, Evaluation of Water Saturation in Shaly Formations: Soc. of Professional Well Log Analysts, 12th Annual Logging Symposium Trans., paper O.

Schlumberger, 1987, Log Interpretation Principles/Applications: Houston, Texas, Schlumberger Well Services Inc.

Schlumberger, 1969, Log Interpretation/Charts: Houston, Texas, Schlumberger Well Services Inc.

Simandoux, P., 1963, Measures dielectriques en milieu poreux, application a mesure des saturations en eau: Etude du Comportement des Massifs Argileux, Revue de L'institut Francais du Petrole, Supplementary Issue.

Sneider, R. M., 1987, Practical petrophysics for exploration and development: R. M. Sneider Exploration, Inc., Houston, Texas.

Tepper, Roy, 1989, A clay corrected neutron-density crossover method as a useful gas/oil interpretation technique for Gulf of Mexico shaly sandstone reservoirs: Soc. of Professional Well Log Analysts, 30th Annual Logging Symposium Trans., paper R.

Tyler, Noel, 1988, New oil from old fields: Geotimes, v. 7, no. 7, p. 8-10.

Vajnar, E. A., C. M. Kidwell, and R. A. Haley, 1977, Surprising productivity from low-resistivity sands: Soc. of Professional Well Log Analysts, 18th Annual Logging Symposium Trans., paper EE.

Visser, R., K. A. T. Baur, and J. P. von Baaren, 1988, Effective porosity estimation in the presence of dispersed clay: Soc. of Professional Well Log Analysts, 29th Annual Logging Symposium Trans., Paper BB.

Waxman, M. H. and L. J. M. Smits, 1968, Electrical conductivities in oil bearing shaly sands: Jour. of Petroleum Technology, (June), p. 107-122.

Wilson, M. D. and R. W. Tillman, 1974, Diagenetic destruction of feldspar and genesis of clays: their influence on sandstone classification and grain size analysis: Geol. Soc. America [abstract] with Programs, v. 5, p. 130-131.

NOMENCLATURE

SP - spontaneous potential log
GR - gamma ray log
SFL - spherically focused log
ILM - medium induction log
ILD - deep induction log
CNL - compensated neutron log
FDC - formation density compensated log
ρ_b - bulk density of formation
EPT - electromagnetic propagation tool

ϕ_{PHIA} - neutron density porosity from computer log
ϕ_{EMC} - EPT porosity from computer log (water-filled porosity)
ϕ_n - neutron porosity uncorrected for clay
ϕ_{nd} - neutron-density porosity uncorrected for clay
ϕ_{nc} - neutron porosity corrected for clay
ϕ_{dc} - density porosity corrected for clay
ϕ_{xp} - cross plot neutron density porosity uncorrected for clay
ϕ_s - sonic porosity uncorrected for clay
ϕ_d - density porosity uncorrected for clay
V_{sh} - volume of shale
I_{GR} - gamma ray index
ϕ_e - effective clay-corrected porosity
S_b - bound water saturation
V_{cl} - volume of clay minerals
S_{wt} - total water saturation
S_{we} - effective water saturation corrected for clay
q - shaliness factor
R_w - formation water resistivity
R_b - bound water resistivity
R_{sh} - resistivity of shale
CEC - cation exchange capacity in meq/gm
ϕ_{tsh} - total porosity of shale
Δt - interval transit time of formation
Δt_{ma} - interval transit time of matrix
Δt_f - interval transit time of fluid
ρ_{ma} - matrix density
ρ_f - fluid density
Δt_{sh} - interval transit time of shale
ρ_{sh} - shale density
S_{wirr} - irreducible water saturation
R_z - mixed water resistivity
R_{mf} - resistivity of mud filtrate
δ - correction factor for ϕ_{tsh} in the Dual Water model
R_{wa} - apparent formation water resistivity
ϕ_t - total non-clay corrected porosity
ϕ_{nsh} - neutron porosity of shale
ϕ_{dsh} - density porosity of shale
SSP - static spontaneous potential
PSP - pseudostatic spontaneous potential
GR_{max} - gamma ray maximum (shale)
GR_{min} - gamma ray minimum (clean sand)
GR_{log} - gamma ray of shaly sand
R_o - wet resistivity
F - formation factor
R_{cl} - dispersed clay resistivity
Q - CEC expressed as milliequivalents/unit volume of pore fluid
B - specific counterion conductivity in mho/m per meq/cc
Counterions - ions in exchangeable sites
ϕ_{nma} - neutron porosity of matrix

APPENDIX I

Additional Shaly Sand Formulas

Waxman - Smit (modified by Hilchie, 1982)

$S_{we} = \{ -BQ_vR_{w2} + [(BQ_vR_{w2})^2 + (4\,R_{wl}/\phi_e^2R_t)\,]^{1/2}\}/2$

for use in low V_{cl} ($V_{cl} \leq 10\%$) shaly sands and shaly sands with brackish formation waters

Poupon (1971)

$S_{we} = \{(R_{sh}\ R_w)^{1/2}/[(R_w)^{1/2}V_{cl}^{\,d} + \phi_t(R_{sh})^{1/2}]\} \times [\,1/(R_t)^{1/2}]$

for use in Indonesian high V_{cl} ($V_{cl} \geq 30\%$) shaly sands with brackish formation waters

Modified VOLAN (Tepper, 1989), a dual water type model

$S_{we} = (S_{wt} - S_b)/(1 - S_b)$

$S_{wt} = \{-S_b(R_w/R_b - 1.0) + \{[S_b(R_w/R_b - 1.0)]^2 + 4(\,R_w/\phi_t^2R_t)\}^{1/2}\}/2$

$S_b = V_{cl} \times [(R_w/R_t)^{1/2}/\phi_t]^{1/2}$

$R_b = R_{sh} \times \phi_{tsh}^2$

where:

S_{we} = effective water saturation
R_t = true formation resistivity
ϕ_e = effective porosity
ϕ_t = total porosity
$d = 1.0 - (V_{c1}/2)$
R_{W_1} = formation water resistivity at formation temperature
R_{W_2} = formation water resistivity at 77°F
$B = 4.6[1.0 - 0.6e^{(-0.77/RW_1)}]$
$Q_v = [CEC(1.0 - \phi_{tsh})\rho]/100\ \phi_{tsh}$
CEC = cation exchange capacity in meq/100 gm
ρ = grain density of clay in gm/cc
V_{cl} = volume of clay
R_{sh} = shale resistivity
S_b = bound water saturation
R_b = bound water resistivity
ϕ_{tsh} = total porosity of shale (see: Dual water method Chapter 3)

APPENDIX II

Additional Methods to Determine Volume of Clay (V_{cl})

1.) DENSITY-SONIC:

$V_{cl} = [\rho_b \times (\Delta t - \Delta t_f)\,] - [\Delta t \times (\rho_{ma} - \rho_f)\,] - (\rho_f \times \Delta t_{ma}) + (\rho_{ma} \times \Delta t_f)\ /\ [\ (\Delta t_{ma} - \Delta t_f\,) \times (\rho_{sh} - \rho_f\,)\,] - [\ (\rho_{ma} - \rho_f\,) \times (\Delta t_{sh} - \Delta t_f)]$

CONDITIONS: Less dependent of lithology and fluid conditions than neutron-density method. (Dresser Atlas, 1983)

2.) NEUTRON-SONIC:

$V_{cl} = [\phi_n \times (\Delta t_{ma} - \Delta t_f)\,] - [\Delta t \times (\phi_{nma} - 1.0)\,] - \Delta t_{ma} + (\phi_{nma} \times \Delta t_f)\ /\ [\ (\Delta t_{ma} - \Delta t_f\,) \times (\phi_{nsh} - 1.0\,)\,] - [\ (\phi_{nma} - 1.0\,) \times (\Delta t_{sh} - \Delta t_f)]$

CONDITIONS:
Use only in gas zones with low S_w. (Dresser Atlas, 1983)

3.) BULK DENSITY:

$V_{cl} = I_{GR} \times (\rho_b/\rho_{sh})^3$

CONDITIONS: General V_{cl} formula for use when there is authigenic clay in pores. (Dewan, 1983)

4.) RESISTIVITY:

$V_{cl} = R_{sh}/R_t$

CONDITIONS: Will overestimate V_{cl} if dispersed clay is present. When dispersed clay is present use $R_{cl} = 0.4\ R_{sh}$ in place of R_{sh}.